OCEANOGRAPHY AND OCEAN ENGINEERING

THE STRAIT OF GIBRALTAR: A FIELD LABORATORY TO ANALYZE BIOLOGICAL RESPONSE TO PHYSICAL FORCING

OCEANOGRAPHY AND OCEAN ENGINEERING

Additional books in this series can be found on Nova's website under the Series tab.

Additional books in this series can be found on Nova's website under the E-books tab.

OCEANOGRAPHY AND OCEAN ENGINEERING

THE STRAIT OF GIBRALTAR: A FIELD LABORATORY TO ANALYZE BIOLOGICAL RESPONSE TO PHYSICAL FORCING

D. MACÍAS
F. ECHEVARRÍA
C.M. GARCÍA
AND
M. BRUNO

Nova Science Publishers, Inc.
New York

Copyright © 2010 by Nova Science Publishers, Inc.

All rights reserved. No part of this book may be reproduced, stored in a retrieval system or transmitted in any form or by any means: electronic, electrostatic, magnetic, tape, mechanical photocopying, recording or otherwise without the written permission of the Publisher.

For permission to use material from this book please contact us:
Telephone 631-231-7269; Fax 631-231-8175
Web Site: http://www.novapublishers.com

NOTICE TO THE READER

The Publisher has taken reasonable care in the preparation of this book, but makes no expressed or implied warranty of any kind and assumes no responsibility for any errors or omissions. No liability is assumed for incidental or consequential damages in connection with or arising out of information contained in this book. The Publisher shall not be liable for any special, consequential, or exemplary damages resulting, in whole or in part, from the readers' use of, or reliance upon, this material.

Independent verification should be sought for any data, advice or recommendations contained in this book. In addition, no responsibility is assumed by the publisher for any injury and/or damage to persons or property arising from any methods, products, instructions, ideas or otherwise contained in this publication.

This publication is designed to provide accurate and authoritative information with regard to the subject matter covered herein. It is sold with the clear understanding that the Publisher is not engaged in rendering legal or any other professional services. If legal or any other expert assistance is required, the services of a competent person should be sought. FROM A DECLARATION OF PARTICIPANTS JOINTLY ADOPTED BY A COMMITTEE OF THE AMERICAN BAR ASSOCIATION AND A COMMITTEE OF PUBLISHERS.

LIBRARY OF CONGRESS CATALOGING-IN-PUBLICATION DATA

Available upon Request
ISBN: 978-1-61728-973-6

Published by Nova Science Publishers, Inc. ✢ *New York*

CONTENTS

Preface		vii
Chapter 1	Introduction	1
Chapter 2	Flows, Tides and Associated Hydrology in the Strait of Gibraltar	5
Chapter 3	Along-Strait Dynamics. Mixing and Advection	11
Chapter 4	Hypotheses, Evidences and Proofs of the Tidally-Induced Coastal-Channel Interactions in the Strait	37
Chapter 5	Conclusion	53
References		55
Index		65

PREFACE

The Strait of Gibraltar is the unique and narrow connection between the Mediterranean basin and the open Atlantic Ocean. One of the main features of this place as the strong topographic constriction that happens both in the horizontal (minimum width of 14 km) and vertical (minimum depth of 250 meters) dimension in conjunction with the presence of a two-layered counter-current circulation and strong inequalities of the tidal range at both sides. These differences in tidal amplitude induce intense barotropic and baroclinic currents which interact with the sharp sea-bottom topography in the main sill of the Strait to create internal waves in the Atlantic-Mediterranean Interface (AMI). These intense and periodic undulatory processes created in the western side of the Strait modify the main along-strait circulation in a number of different ways. On the one hand, they are able to mix the Atlantic and Mediterranean waters creating pulsating upwelling events which greatly alter the biogeochemical properties of the incoming surface layer. Also, during the formation of the train of internal waves over the Sill there is a creation of alternating bands with different vertical and horizontal velocities (associated to the underlying troughs and crests of the waves) that could accumulate biological material in narrow bands across the main channel of the Strait. Finally, they could also generate horizontal divergences being responsible for the suction of coastal waters which are then transported to the inner Mediterranean basin along with the main flux.

All these processes acting together significantly influence the biogeochemical budget of the entire Mediterranean basin, so a great research effort has been made in recent years to elucidate the fundamental mechanisms controlling these hydrological processes and how they control and shape the biogeochemical patterns in the area. In this chapter a comprehensive review of the most recent findings is presented along with some directions for future research necessary to enhance our knowledge of the tidal influence on the inter-basin biogeochemical budget.

Chapter 1

INTRODUCTION

The Strait of Gibraltar is quite a unique place from various points of view including the social, economic and oceanographic dimensions. From an anthropogenic perspective, during a long time it was considered the end of the known world and, even today it is perceived as a natural border between the developed European continent and the developing Africa. It is also a particular place for migrating species as it constitutes the natural pathway for multiple animals including birds, fishes and marine mammals moving across continents or between marine basins.

From a purely geographic perspective, the Strait of Gibraltar is the Mediterranean's only communication with the world ocean, hence its significance. It extends about 60 kilometres in a southwest-northeast, 15 km wide at its narrowest section (Tarifa narrows, red line in Figure 1) and only 280m deep at its main sill (Figure 1). Its reduced dimensions, relatively easy accessibility and strong physical forcing make possible to consider the Strait as a *field laboratory* for oceanographic research.

Coastal topography is quite complicated in the Strait as it presents both capes and bights with sharp latitudinal and longitudinal gradients, which interact with the main currents creating a very complex and variable circulation pattern. Bottom topography highlights the presence of two sills. The main sill, known as Camarinal Sill, is very shallow and defines the smallest section that plays like a bottleneck for water exchange between the basins. The second, Spartel Sill, is deeper and placed to the west on the channel that runs in a southerly direction along Majuan ridge by which take place the main part of the Mediterranean water drainage. Between the two sills is located the Tangier basin (with a maximum depth of over 600 m), which is a smal reservoir of significant importance for tidal dynamics. To the east of Camarinal Sill, the bottom falls in a sharp way to 900 m in the eastern side of the Strait (Figure 1).

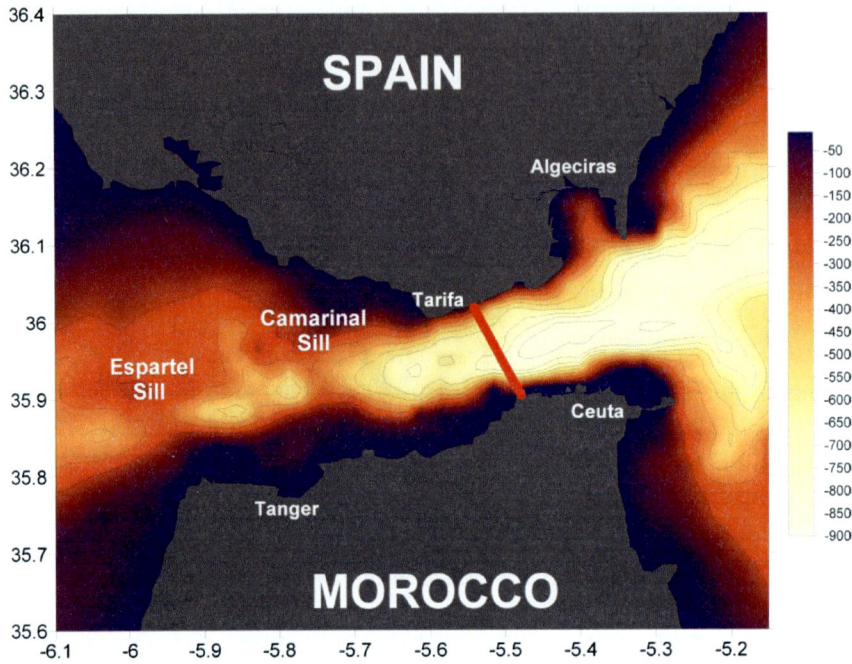

Figure 1. Strait of Gibraltar. Bottom topography from 900 meters each 100 meters. Main cities and bottom signatures are shown. Red line marks the position of the Tarifa Narrows.

The mean water circulation through the Strait of Gibraltar is mainly created and maintained by the negative hydrological budget of the Mediterranean Sea where evaporative losses exceed the freshwater income by both riverine discharges and direct rainfall. These hydrological conditions induce a well-known inverse estuarine circulation along the main channel of the Strait (Lacombe and Richez, 1982; Armi and Farmer, 1988; Hopkins, 1999). This circulation is basically composed by a surface eastward flow of nutrient-poor, open-ocean Atlantic waters and a deep outflow (i.e. westward) of nutrient-rich Mediterranean waters which leads to a natural tendency to oligotrophy in the whole Mediterranean basin. The biogeochemical budget of the Mediterranean Sea depends on the water exchange through the Strait of Gibraltar, as well as on atmospheric and river inputs (Béthoux et al, 1998).

Thereby, the study of the balance of water and elements trough the Strait as well as its dynamics, has implications not only at the regional level but, also, for large basin scale budget calculations (Packard et al, 1988; Minas et al, 1991).

Over annual timescales the water exchange through the Strait of Gibraltar can be regarded as a nearly constant inflow of Atlantic waters towards the Mediterranean in the upper layer (Atlantic Jet, AJ thereinafter) and a nearly constant outflow of deep Mediterranean waters toward the Atlantic beneath (i.e. the antiestuarine circulation presented above). The magnitude of these virtually regular flows has been estimated by direct measurement (e.g. Bryden and Kinder, 1988; Pettigrew, 1989; Bryden et al, 1994; García-Lafuente et al, 2000; Tsimplis and Bryden, 2000) and by numerical models (Wu and Haines, 1996; Sein et al, 1998; Hopkins, 1999; Sanino et al, 2002) and are considered to be basically dependent on the climatic conditions over the Mediterranean area and of the Strait's main geographical characteristics (Bryden and Kinder, 1991; García Lafuente and Criado, 2001).

However, this description of the along-strait circulation is a simplification of the real one given that at least three different water masses can be observed in the region: Surface Atlantic Water (SAW), North Atlantic Central Water (NACW) (which together constitute the AJ) and Mediterranean Outflowing Water (MOW) (Gascard and Richez, 1985), the amount of each one being strongly dependent on multiple factors including the tidal height (Gascard and Richez, 1985) and the along-strait position (Bray et al, 1995).

Another consequence of the tidally-intensified currents comes from their interaction with the bottom topography. As stated above, bottom depth changes dramatically near the Camarinal Sill (Figure 1). The interaction between flows and this sharp topography leads the Atlantic Mediterranean Interface (AMI) to changes its position abruptly too. This is particularly true during certain phases of the tidal cycle, when the abrupt changes of the AMI are related either to the formation of internal hydraulic jumps (Boyce, 1975; Armi and Farmer, 1985), a phenomenon that prevails during moderate to strong (spring) tides, or arrested internal waves (Bruno et al, 2002), which are more usual during weak (neap) tides. Such undulatory processes enhance interfacial mixing (Wesson and Gregg, 1994) and can inject deep, nutrient-rich water into the upper layer of Atlantic water. The upwelled inorganic nutrients are advected towards the Mediterranean Sea in the upper layer, enhancing, as a result, the primary production in the Alboran Sea to the east of the Strait. The turbulence-favouring abrupt nature of hydraulic jumps relative to the smoother arrested waves suggests a fortnightly cycle for mixing and, hence, for the exchange of dissolved substances. The simultaneous occurrence of enhanced mixing and strong tidal currents can give rise to positive correlations between nutrient concentration and tidal flows, in which case, the mean flux into the Mediterranean estimated above by assuming steady flows needs to be revised.

Also, water composition of the AJ has been described to be influenced by the tidal forcing. As an example, the quantity of the less-abundant NACW is clearly linked with tidal amplitude (Gascard and Richez, 1985, Macías et al, 2006) or wind regime (Gómez et al, 2004). Water mass distribution in the AJ and biological patterns are highly related in this region (Gómez et al., 2001; Macías et al., 2006, 2008) characterised by a discontinuous [horizontal entrainment] transport of chlorophyll patches into the Alborán Sea.

The vertical distribution of chlorophyll is also expected to be linked to the dynamics of the water masses, which show clear vertical segregation, with SAW at the surface, MOW in the deeper layers and NACW at mid-depths. Thus, three important contact zones can be defined, SAW-NACW, SAW-MOW, and NACW-MOW, where according to previous observations (Macías et al., 2006) using *in vivo* chl *a* fluorescence, deep chlorophyll maximum (DCM) are likely to occur. Those DCMs have been also characterised by their biogeochemical signature (Macías et al., 2008) with a great diversity associated also with the tidal intensity. These differences in composition of the DCM according to their origin and position could be of particular relevance for the pelagic ecosystem of the Alboran Sea as the discontinuous entrance of biogeochemical material should affect the composition and behaviour of the planktonic community.

In the present book we will (1) describe the main characteristics of the flow and tidal dynamics within the Strait of Gibraltar and in the nearby areas of both the Atlantic Ocean and the Mediterranean Sea. Afterwards we will present some of the most prominent biogeochemical signatures in the region associated with the tidal forcing including (2) along-strait dynamics associated with interfacial mixing events, (3) diversity and dynamics of the planktonic community composition in the main channel and, finally, (4) coastal-channel interactions in the Camarinal Sill region and associated effects on the biogeochemical budget between basins.

Chapter 2

FLOWS, TIDES AND ASSOCIATED HYDROLOGY IN THE STRAIT OF GIBRALTAR

As already pointed out in the introduction, the water flux through the Strait of Gibraltar is quite complex being possible to identify at least four main components (Lacombe and Richez, 1982; Candela, 1991): a tidal, mainly barotropic flow, with magnitudes up to $2.5 ms^{-1}$ (Candela et al, 1990); a barotropic subinertial component (with periodicity ranging from days to several months) driven by atmospheric pressure fluctuations within the Mediterranean and with magnitudes close to $0.4 ms^{-1}$ (Candela et al, 1989); a long-term baroclinic component driven by the internal pressure gradient due to the density difference between the Mediterranean and the Atlantic Waters, with magnitudes of about $0.5\ ms^{-1}$ (Bryden et al, 1994); and subtidal currents associated to large amplitude internal waves induced by the interaction of tidal flows with vertical stratification and bottom topography which are mainly generated around Camarinal Sill (Armi and Farmer, 1988; Bruno et al, 2002; Vázquez et al, 2006).

Therefore, from a physical point of view the Strait is a very energetic system with long-term, subinertial, tidal and subtidal currents all being of significant amplitude. In this second chapter of the book each one of these temporal scales of the flux will be presented and analysed.

2.1. LONG-TERM DYNAMICS

Long-term currents are generally characterised by a two-layer flow with the Atlantic water flowing in the upper layer towards the Mediterranean and the Mediterranean water doing it in the lower layer towards the Atlantic. The reason for this two-layer exchange is to compensate the negative budget of water due to strong evaporation in the Mediterranean Sea. The upper layer flow (Atlantic inflow) is driven by a sea level difference between western and eastern sides of the strait of about 15 cm (García-Lafuente, 2008) while the lower layer flow (Mediterranean outflow) is driven by the density difference between the two basins.

Surface Atlantic Water (SAW) and, to a lesser extent, North Atlantic Central Water (NACW), whose salinity varies between 36 and 36.5 compose the Atlantic inflow that originates in the Gulf of Cádiz. Most Mediterranean outflow consists of Levantine Intermediate Water (LIW), which is formed in winter in the eastern Mediterranean (Rhodes Basin) and returns as an intermediate counter-current through the Strait of Sicily (Pettigrew, 1989), reaching the Strait of Gibraltar once it skirts the European coast in an anti-clockwise sense. Western Mediterranean Deep Water (WMDW) is the second type of outflowing water and forms in winter in the Gulf of Lion. It is colder, less saline and slightly denser than LIW and is located beneath it. The volume of WMDW formed depends on the harshness of winter. Consequently it shows a high inter-annual variability in the Strait (García Lafuente et al, 2007). Due to the strong vertical mixing affecting to LIW and WMDW they are indistinguishable when they cross the Strait towards the Gulf of Cadiz.

In the Strait, salinity is the variable that separates the upper and lower layer flows. It is usual to take the surface of 37.5 as a reference. The time averaged position of this surface, commonly known as interface, slopes up along the axis of the Strait from 200-250 m offshore of Cape Spartel to less than 100 m at the eastern boundary. There is also a cross-strait slope induced by the Earth's rotation that raises the mean position of the interface from south to north.

Long-term flows through the Strait show a clear seasonal variability that is clearly linked to the formation of WMDW (García-Lafuente, 2008), as it affects the outflow characteristics, being slightly increased and colder in late spring and diminishing at the end of the year. Furthermore, seasonal warming of surface waters, which causes a marked density contrast in summer, enhances the Atlantic inflow in late summer.

2.2. SUBINERTIAL DYNAMICS

On subinertial time-scale (days to a few months), the variation in atmospheric pressure over the Mediterranean basin is the main cause of fluctuations in the current intensity through the Strait. The physical mechanism is related to the isostatic response of sea level variations to the atmospheric pressure fluctuations over the western Mediterranean (Crepon, 1965; Candela et al, 1989; García-Lafuente et al, 2002). Also, subinertial flows are almost 180° out of phase with respect to the atmospheric pressure fluctuations for oscillation periods ranging from 3 to 80 days. In addition to these meteorologically forced flows, there is a significant part of the subinertial fluctuations, showing a baroclinic behaviour with opposite current directions below and above the interface, which is responsible for the temporal variations of the vertical shear of currents, that shows its maximum/minimum during neap/spring tides. Bryden et al (1994) pointed out that variation in tidal mixing intensity between spring and neap tides cycles might be a likely mechanism to explain this behaviour.

2.3. TIDAL DYNAMICS

It is customary in tidal analysis to distinguish between barotropic and baroclinic components of the tide. The barotropic part is characterised by the vertical displacement of the sea level (surface tide) and the currents forced by the horizontal gradient. On the other hand, the baroclinic tide is characterised by the vertical displacement of the isopicnal surfaces (internal tide) and the currents originated by the induced horizontal gradients of density.

Concerning the barotropic or surface tide, the Strait of Gibraltar links the large tidal range of the Atlantic Ocean (exceeding 3 m during spring tides) with that of the Mediterranean, where tides are, in general, practically non-existent. In the Strait the dominant oscillations are of semidiurnal period, with amplitude of 1 m at the western side decreasing down to 0.3 m at the eastern side (Candela et al, 1990). Once inside the Mediterranean it becomes null in Alicante at the eastern end of the Alboran Sea basin.

Barotropic tidal currents are directed towards the Atlantic between low and high tide, carrying the water masses demanded to adjust the level of the oceanic high tide. During the ebb the tidal currents flow towards the

Mediterranean, draining water to adjust the low tide in the oceanic side of the Strait.

Since the amplitude of the tidal currents is considerably greater than the mean currents, it would be expected a periodical reversion of the current direction, at all depths of the water column. However, it does not occur at all places along the Strait due to the interaction of the intense tidal flow with the abrupt bottom topography and the strong vertical stratification of the water column, which originates an important internal tide. As a result of this interaction, the surface current in the easternmost part of the Strait is never reversed by tidal currents and the same occurs with the Mediterranean outflow in the westernmost area.

2.4. SUBTIDAL DYNAMICS.
LARGE AMPLITUDE INTERNAL WAVES

The interaction of the barotropic tidal flow with the main sill (Camarinal Sill) topography and the stratified water column, primarily causes internal tides that evolve, by non-linear processes and non-hydrostatic dispersion, into large-amplitude (more than 100m) internal waves exhibiting much shorter period and wavelength than those related to the basic tidal variability (Richez, 1994; Bruno et al, 2002).

As discussed by Bruno et al (2002), once internal waves are formed they are trapped on the lee side of the sill because of the establishment of critical or supercritical conditions over the sill and it is an important factor explaining the growth in amplitude that internal waves experience here. In practise, critical/supercritical conditions mean that celerity of the internal undulations is equal/less than a depth-averaged current intensity flowing in the opposite direction. Some of the internal undulations generated around Camarinal Sill during the flooding phase of tidal currents, are also propagated towards the Atlantic. However, they are almost imperceptible because (i) they are rapidly advected towards the Atlantic (having no time to grow) and (ii) the deeper position of the interface west of Camarinal Sill.

The release of the internal waves towards the Mediterranean begins with the establishment of subcritical conditions over the sill (i.e internal wave celerity is greater than depth-averaged current). It happens almost at the beginning of the inflow phase of the barotropic tidal currents (towards the

Mediterranean Sea). In Figure 2 the different stages of the internal wave generation are illustrated.

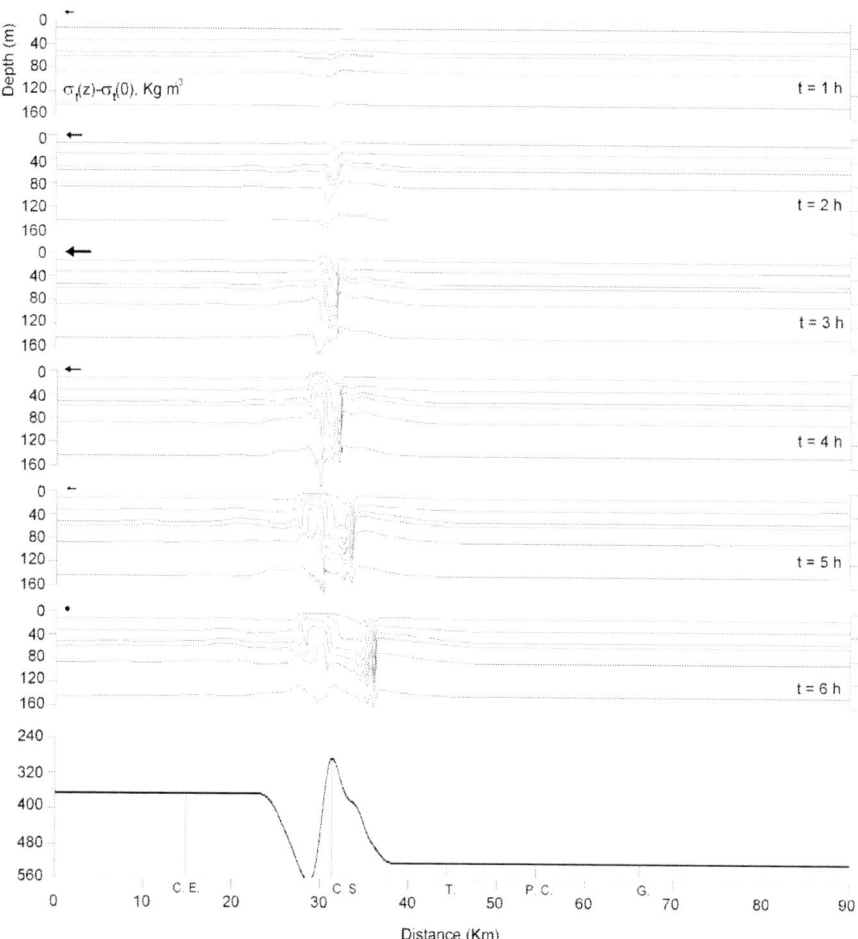

Figure 2. Vertical sections of density due to internal tide as simulated with a non-hydrostatic numerical model in an idealised Strait of Gibraltar. Arrows on the left upper corner of each section indicates the direction and intensity of barotropic tidal currents at Camarinal Sill (CS). At time t=2 h internal tide begins to be generated at CS left flank of the perturbation propagates against the barotropic flow. At time t=3 h perturbation is trapped at Camarinal Sill due to the critical conditions over there and a hydraulic jump (internal bore) is generated. At t= 4 h, hydraulic jump disintegrates into several undulations and finally at t=5 h internal perturbations are released towards the Alboran Sea.

These internal waves have been found to be one of the major contributors to the mixing between the Atlantic and Mediterranean layers within the Strait (Wesson and Gregg, 1994; Macias et al, 2006), being able to have significant remote effects on the hydrography of the Alboran Sea (Vázquez et al, 2006). This fact confers special importance to the study of internal wave phenomena in the Strait of Gibraltar.

Vázquez et al (2008) have inferred on an empirical basis, using a long record of ADCP profiles over Camarinal Sill, that large amplitude internal waves are generated when barotropic tidal currents during flood tide (flow towards the Atlantic) reach an intensity of 1 m/s. Also, the release of the internal waves takes place when intensity is reduced down to 0.5 m/s. This result enables the use of tidal current predictions over Camarinal Sill to forecast the occurrence of large amplitude internal waves around Camarinal Sill. However, the same authors have found that the empirical model fails in some occasions due to the effect of subinertial flows. They concluded that modification of the hydraulic conditions over the sill produced by subinertial flow variations (forced by atmospheric pressure fluctuations in the western Mediterranean) is able to produce an activation/inhibition of internal wave events during neap/ spring tides. This could make possible to activate these internal wave events during neap tides, when predicted hydraulic conditions over the sill do not favour their occurrences, and to inhibit their generation during spring tide conditions, when large internal wave events are expected to occur.

Once internal waves are released they travel towards the Alboran Sea. The time spent to arrive to the Alboran Sea, is significantly affected by the diurnal inequality of tidal currents and also by subinertial flow variations that must exert some effect at a longer time scale. Sánchez-Garrido et al (2008) have reported that due to the diurnal inequality, the travel time of internal waves between Camarinal Sill and the eastern side of the Strait range from 8 to 11 hours.

Chapter 3

ALONG-STRAIT DYNAMICS. MIXING AND ADVECTION

As stated in the previous chapter of the book, the most important and energetic processes in the Strait happens in the west-east direction along its main channel including the two-layer circulation and the intense interfacial mixing (between Mediterranean and Atlantic layers) in the internal-waves generation area (i.e. the Camarinal Sill). Thereby, it seems reasonable to make a first approach to the general dynamics of the area by focusing in the along-strait processes of water flow and mixing events.

Although the reported water flux values across the Strait differ from each other, a reasonable value of general agreement is around 0.8 Sv (1 Sv = 1×10^6 m^3/s), the inflow being around 5% greater than the outflow in order to compensate for evaporative losses in the Mediterranean Sea. These mean fluxes (in and out of the Mediterranean basin) correspond approximately to 300 times the mean flux of the Nile River, which imply that, every second, roughly 600 Nile Rivers are being interchanged through the Strait of Gibraltar.

This long-term average circulation pattern exhibits large fluctuations at different time scales as presented in the previous chapter. Seasonal and subinertial (meteorologically-induced) fluctuations of, typically, 0.1 Sv and 0.5 Sv, respectively have been reported (Candela, 1990; García-Lafuente et al, 2002), but the main source of variability is tidal. There are significant differences in tidal amplitude between the western and eastern parts of the strait, and this induces barotropic and baroclinic tidal currents along the main channel (Lacombe and Richez, 1982) whose amplitude can be up to 4 Sv during spring tides, more than four times greater in magnitude than the time-averaged flow (García-Lafuente and Vargas, 2003).

An interesting and curious fact pointed out by Bryden et al (1994) is that tides contribute to the mean exchange through the positive correlations between the position of the interface separating Mediterranean and Atlantic waters (AMI) and the strength of the tidal currents. Bryden et al (1994) showed that, on average, almost half of the flow exchange measured in the main sill of Camarinal, occurs by virtue of this correlation. The analysis of Vargas et al (2006) confirmed this mechanism (which they named tidal-rectification of flows) but they went further and showed that tidal rectification dominates the exchange during spring tides and is negligible during neap tides. The question remains open as to whether or not the exchange of other substances also follows a pulsating pattern related to the fortnightly cycle of tides, an issue that becomes more complex if mixing is taken into account.

These reported values of interchanged flows have their reflection on the biogeochemical budget between both marine basins. For example the nitrate concentration of the exchanged waters is estimated as 1.2 mmol N/m^3 in the inflowing Atlantic waters and 9.6 mmol N/m^3 in the outflowing Mediterranean waters (Gómez et al, 2000b; Minas et al, 1991; Dafner et al, 2003). With these concentrations and using the flow estimates of Bascheck et al (2001), the nutrient fluxes towards the Mediterranean Sea and towards the Atlantic Ocean would be 972 and 7296 mol N/s, respectively. Thus, the Mediterranean Sea would export a net amount of 6324 mol N/s (or 2914 ton N/year) to the global ocean through the Strait of Gibraltar. But this is only an average calculation, which should be greatly influenced by the intense and quick hydrological processes (mainly tidally-related) happening within the Strait.

Due to the high frequency of such processes it is difficult to study them by using the classic approach of vessel-based field-data sampling because of the time needed to take and process the samples. Thereby, alternative methods should be used in order to adequately describe this kind of quick processes. In this sense two alternative but complementary approaches have been taken during the last half-decade of oceanographic research in the area. On the one hand, a specific-designed boat-based sampled was conducted by performing observation on a fixed position within the main channel of the Strait along several tidal cycles. This sampling strategy allows to register the characteristics of the Atlantic Jet (AJ) coming through the Strait along the tidal cycle and to relate them with the timing and amplitude of the tide.

Another alternative is to use numerical hydrological-biogeochemical coupled models. There have been several attempts to simulate the water circulation through the Strait of Gibraltar using hydrodynamic models (e.g. Wang, 1989 and 1993, Brandt et al, 1996, Sein et al, 1998). These models have different levels of complexity ranging from the simplest one-dimensional two-layer model to 3D coarse resolution models. Recently, high spatial-resolution models allowing for realistic bottom topography-flow interaction have been developed: both two-dimensional, two-layer types (Izquierdo et al, 2001; Castro et al, 2004) and three-dimensional ones (Sannino et al, 2002, 2004). All of them have some degree of success in simulating the short-scale undulatory phenomena over Camarinal Sill but they are either unable to deal with mixing (two-layer models) or mixing was not specifically addressed in the study (Sannino et al, 2004).

However, less attention have been paid to the development of physical-biological coupled models that explore the effect of the strong advection and mixing processes on biogeochemical exchanges and the behavior of the pelagic ecosystem in the Strait and adjacent marine regions. A conceptual model of the plankton distribution in the Strait was proposed by Gómez et al. (2000a) and by Echevarría et al. (2002). These papers related the quasi-permanent enrichment of phytoplankton biomass in the north-eastern area of the Strait to mixing processes over the Camarinal Sill and the subsequent eastwards advection of the water masses. However, these authors did not carry out numerical calculations in order to test this conceptual model.

The circulation and its variability are known to modify the distribution patterns of biological variables in the Strait (Gómez et al, 2001; Macías et al, 2006) and in its neighbour areas (Mercado et al, 2005). Therefore, the correct simulation of the pelagic community in this area must be achieved by means of a physical-biological coupled model in which the physical part can resolve short time-scale dynamics features such as tidal mixing. Physical-biological modelling is actually a fruitful approach promoted by international programs as GLOBEC.

In this framework, a 2-layer 1-D model was developed by Macias et al (2007) by extracting an along-strait W-E section (red line in Figure 3a) from the advanced hydrodynamic model of Izquierdo et al (2001). This 1D model was created to improve the representation of vertical mixing and a biogeochemical model was coupled to the hydrodynamic one to examine the effect of the mixing and advection processes in the main channel of the Strait on the distribution patterns of biogeochemical fields in the region.

In this chapter of the book the predictions of the model developed by Macías et al (2007) will be used to explore the consequences of the tidal dynamics on the biogeochemical budget through the Strait. The validity of this 1D approach will be checked by comparing the model predictions with the field observations taken on the diel cycles at a fixed position in the main channel of the Strait.

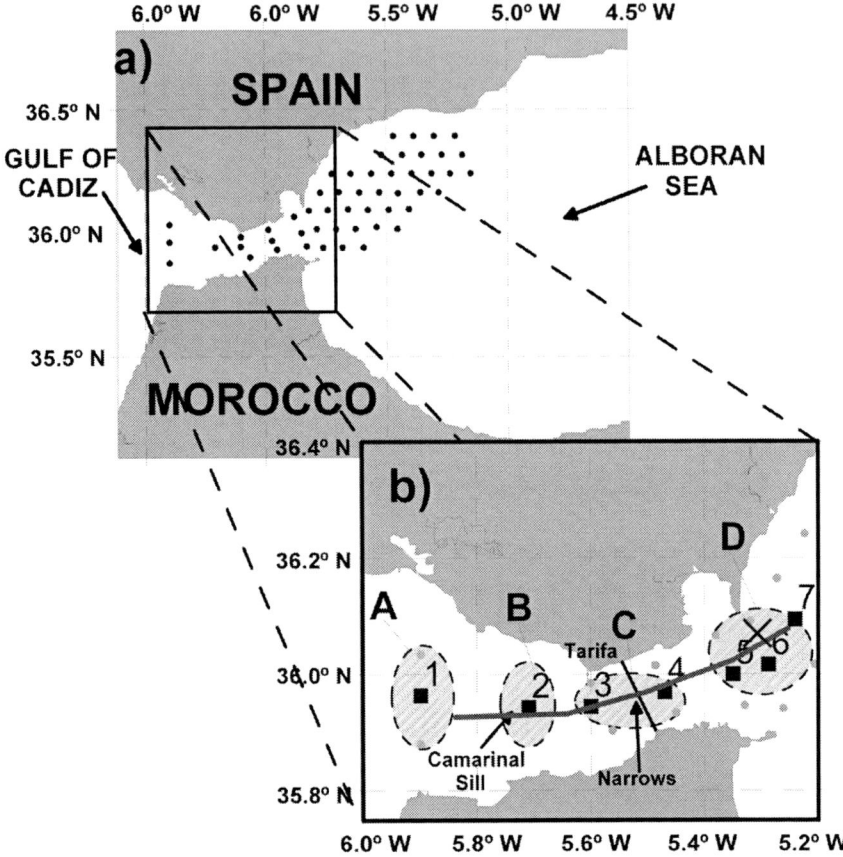

Figure 3. a) Grid of stations sampled in the Strait and NW Alboran Sea. b) Stations selected within the main channel of the Strait to create the conceptual model of water masses circulation and DCM position. Blue line marks the section simulated with the biological-hydrological coupled model and the blue cross indicates the position of the diel sampling used to compare with the model.

However, these two approaches focused on the temporal evolution of the hydrological and biogeochemical signatures in the Strait, sacrificing the vertical resolution of such patterns in order to get the adequate temporal resolution. Nevertheless, it is expected that those processes have also consequences on the vertical distribution of both water masses and pelagic biomass within the Strait.

For example, it is widely recognized that one of the most distinct and ubiquitous biological features in the world's ocean is the presence of a Deep (or sub-surface) Chlorophyll Maximum (DCM) associated with discontinuities in the water column, such as the seasonal thermocline or the presence of haloclines, particularly where different water masses meet. The existence of a DCM, in conjunction with seasonal variations in the depth of the mixed layer, is of crucial importance for explaining the annual cycle of primary production in the open ocean (Gran, 1931; Sverdrup, 1953). Thus, coupled biological-physical studies in regions with marked vertical gradients (such as the Strait itself) should focus on the vertical dimension, since it is presumably in these interfaces, characterized by marked gradients, where phytoplankton cells tend to accumulate (Mann and Lazier, 1991; Rodriguez et al, 1998).

The relationship between DCM, water masses and hydrodynamic conditions has been extensively studied both in the Mediterranean basin and in the Gulf of Cadiz, the two oceanographic regions connected by the Strait of Gibraltar (Estrada et al, 1993; Morán et al, 2001; Navarro et al, 2006) and its distribution within the Strait itself should be far more complex because of the simultaneous presence of the different typical water masses (SAW, NACW and MOW) leading to the presence of multiple density interfaces. This complex dynamics will be described in detail in the final part of this chapter of the book. All this information will serve to propose a conceptual 2D model that represents patterns of water mass circulation and biological variables distribution in relation to the tidal cycle within the Strait, expanding, thereby, the conclusion obtained with the more simplistic 1D approach.

A brief description of the used numerical model and the field data samplings is presented in subsection 3.1; subsection 3.2 focused on describing the 1D-along-strait dynamics based on the diel cycle observations and the model simulations while in subsection 3.3 the conceptual 2D (vertically) model of water circulation and DCM dynamics is presented.

3.1. MODELS AND FIELD DATA SAMPLING DESCRIPTIONS

3.1.1. Hydrodynamic Model

The physical part of the physical-biological model is a 2D, nonlinear, two-layer, free-surface, hydrostatic model with boundary-fitted curvilinear coordinates. Sea-water density is uniform and prescribed in each layer. A complete model description, including governing equations and parameter values used can be found in Izquierdo et al (2001). From the complete grid of the hydrodynamic model, an along-strait section was selected (red line in Figure 3a) situated in the centre of the main channel of the Strait.

The model is forced at the open boundaries with radiation-type boundary conditions ensuring that when short-wavelength disturbances in the fields of variables are generated they all propagate away from the region of interest.

At the coastal boundaries a condition of null normal flow is applied. In order to reduce the influence of any inaccuracies in boundary forcing upon the sought-for solution the waves produced within the strait are allowed to propagate freely through its open boundaries. The M_2, S_2, K_1 and O_1 surface tidal elevation amplitudes and phases used to set the tidal forcing at the open boundary grid points were derived by interpolating the relevant values from a 0.5° gridded version of the FES95.2 global tidal solutions of Le Provost et al (1998). The model was run for 30 identical semidiurnal tidal cycles to achieve a stable time-periodic solution. After establishing this solution, the model run was continued for a 13-month period.

3.1.1.1. Mixing-Advection Model

The physical processes influence the biogeochemistry of the region through mixing and advection. Advection velocities for all tracers are provided directly by the model output. Mixing, which is responsible for variations of concentration, is not directly computed from the hydrodynamic model, which is immiscible and therefore does not allow for any exchange of properties between the layers. However, taking into account that interfacial mixing is strongly dependent on the vertical velocity shear across the interface (Briscoe, 1994) a parameterisation scheme to estimate interfacial mixing was developed. The success of this parameterisation is supported by the good reliability of the tidal current predictions of the used hydrodynamic model (Brandt et al, 1996).

3.1.1.2. Biogeochemical Model

The biological model is a simple nitrogen-based Nutrient-Phytoplankton-Zooplankton (NPZ) model. To agree with the numerical grid of the hydrodynamical model for the along-strait transect, the upper layer were divided into fixed volume cells with the same along-strait length.

Boundary and initial conditions for the concentration of nutrients, phytoplankton and zooplankton come from the analysis of more than 150 field data taken all over the Strait at five different depths during four different cruises carried out in several years being in good agreement with previous observations in the region (Gómez et al, 2000b; Minas et al, 2001; Dafner et al, 2003).

An NPZ model is a very simple description of the pelagic plankton ecosystem. A more complex model, like that described by Fasham et al (1990) with seven different compartments including bacteria and detritus, was also tested but the differences in model performance were very small for all N, P and Z.

3.1.2. Diel Cycle Sampling

Data were collected during four different cruises carried out from November 2002 until November 2003 in the Strait of Gibraltar. In each cruise one or two fixed stations were sampled in the eastern side of the Strait (blue cross in Figure 3b).

In each fixed station (3 to 26 h total observation times) several CTD profiles were made following an interval from 0.25 hours to 1 hour. In each profile, salinity, temperature and fluorescence distribution were sampled from surface to 300 meters depth by using a combined CTD probe. This limit of depth was chosen to ensure observation of the Atlantic-Mediterranean-Interface (AMI) and MOW but increasing time resolution. The dates of the fixed stations were selected to include different tidal amplitude cycles, from spring to neap tides, with associated along-strait current velocities ranging from 2.5 m/s to 0.7 m/s.

3.1.3. DCM Sampling

Data were obtained during two simultaneous cruises in November 2003 performed in the area of the Strait of Gibraltar and western Alboran Sea on board two research vessels, BIO *Hespérides* and BO *Mytilus*. A grid that covered the studied area (Figure 3a) was surveyed twice at different tidal amplitudes (spring and neap tides) and wind regimes (westerlies and easterlies). From the complete grid, a set of seven stations that characterised the main hydrographic features of the Strait was selected (Figure 3b). Station 1 was used to illustrate the characteristics of the water column on the western side of the Strait (region A, Figure 3b). Station 2 described the water column on the main sill of the Strait (region B, Figure 3b). The vertical distribution of variables in the central section of the channel (region C, Figure 3b) was examined using the casts performed in stations 3 and 4. The description and monitoring of the eastern sector of the Strait (region D, Figure 3b) was based on data acquired at stations 5, 6 and 7. Each station was sampled several times making a total of 32 observations within the main channel of the Strait.

Basic information on physical structure of the water column at each station was obtained from CTD casts while discrete water samples were also collected at different depths to analyse several variables related to the dynamics of the pelagic ecosystem. Each observation was classified according to the specific period of the tidal cycle to generate the conceptual model of vertical distribution of water masses and biogeochemical properties along the main channel of the Strait.

3.2. 1D (HORIZONTALLY) ALONG-STRAIT DYNAMICS

3.2.1. Field Data

Tidal induced dynamics can be studied by analysing the time evolution of basic descriptive variables (temperature, salinity, chlorophyll fluorescence and tidal current speed) recorded during the seven samplings events (Figure 4). The details about the prediction of the tidal current shown in plate d could be revised in Alonso et al (2003). This figure is but an example of the seven sampling performed on the eastern entrance of the main channel of the Strait (see Figure 3a) and it could be used to track the presence and vertical position of the different water masses which appears in the Strait.

Each CTD cast was classified into two different types; when NACW was clearly present it was assumed that no significant interfacial mixing (between Atlantic and Mediterranean waters) have taken place, these cast were, thereby, classified as "not mixed". On the other hand, if no NACW was detected the TS diagram (see details in Macías et al., 2006) depicted a progressive gradient from surface SAW to deep MOW, indicating some degree of interfacial mixing, being then classified as "mixed". These two types of water masses appear clearly separated in time during each cycle at the fixed stations. The presence of mixed waters is usually detected around HW-4 (i.e. 4 hours before high-water), when the current velocity over the Sill changes from inflowing (positive values in plate d Figure 4) to outflowing (negative values in plate d Figure 4).

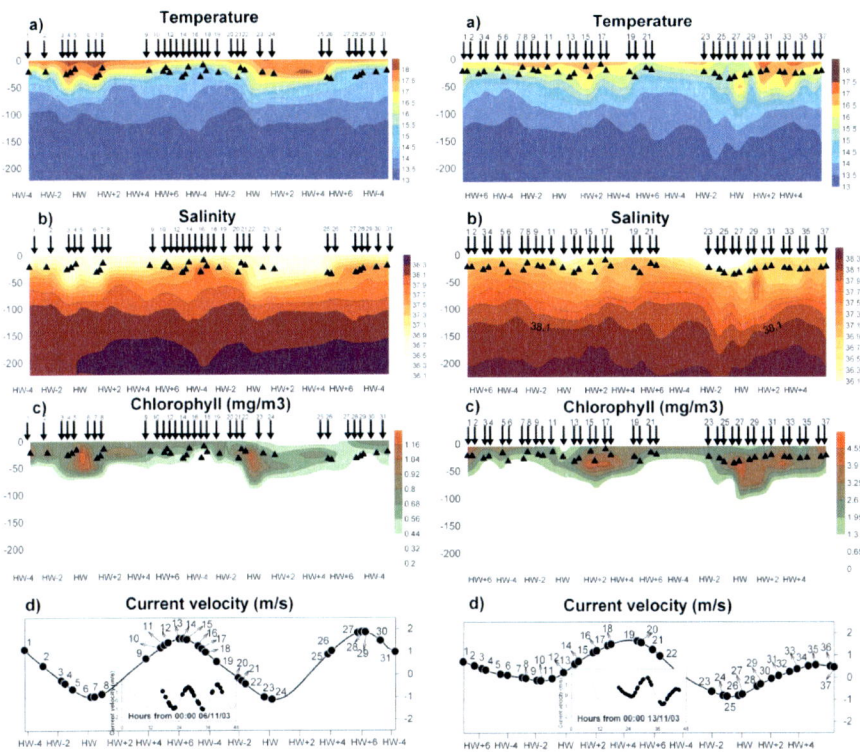

Figure 4. Temporal evolution of salinity (p.s.u.), temperature (°C), Chlorophyll (mg/m^3) and current velocity prediction over the Camarinal Sill (m/s) during the diel cycle sampling. Squares showed the presence of mixing of AMW and the triangles the presence of MMW. The arrows and circles mark the moments when CTD profiles were made.

Several undulatory phenomena were detected during the different sampling events, generally associated with the presence of such mixed waters (Figure 4). These disturbances are marked by vertical displacement of isotherms and isohalines and can be related with the internal waves generated over the Camarinal Sill, travelling within the main flux towards the Mediterranean.

Another common pattern observed is that the highest levels of chlorophyll coincide with the presence of these mixed waters and internal waves (Figure 4), suggesting a strong coupling of the physical forcing and the biological patterns that will be discussed later. Actually, mean chlorophyll concentrations in the mixed waters were in average 28% higher than in the Atlantic waters

(Figure 5) independently of the sampling date and of the absolute value of chlorophyll concentration.

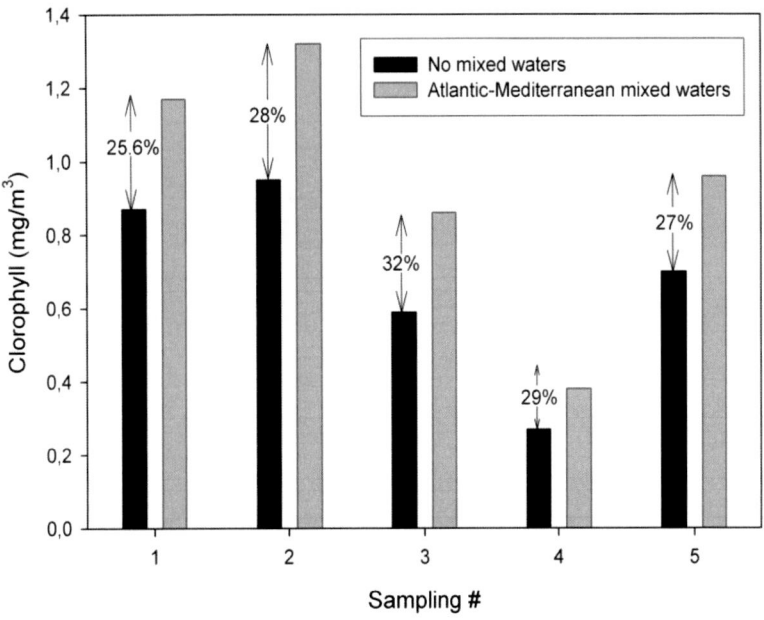

Figure 5. Mean chlorophyll concentration in the upper 75 meters of the water column during the different sampling in the fixed station. Black bars correspond to moments of no mixed water column and grey bars to mixed waters.

3.2.2. Model Simulations – Field Data Comparison

In order to evaluate the validity of the mixing and advection schemes adopted in the model a non-reactive tracer such as salinity was firstly chosen to be compared with the field data, as the temporal evolution of such variable should not be affected by the biogeochemical simplifications assumed by the model. Figure 6 shows the salinity in the upper layer predicted by the model (blue line) and the actual data taken from the field data sampling presented above (black dots) along 24 hours during a spring-tide period. The observed and predicted values were in the same range and the temporal dynamics is very well reproduced by the model (see Macías et al., 2007 and 2007a).

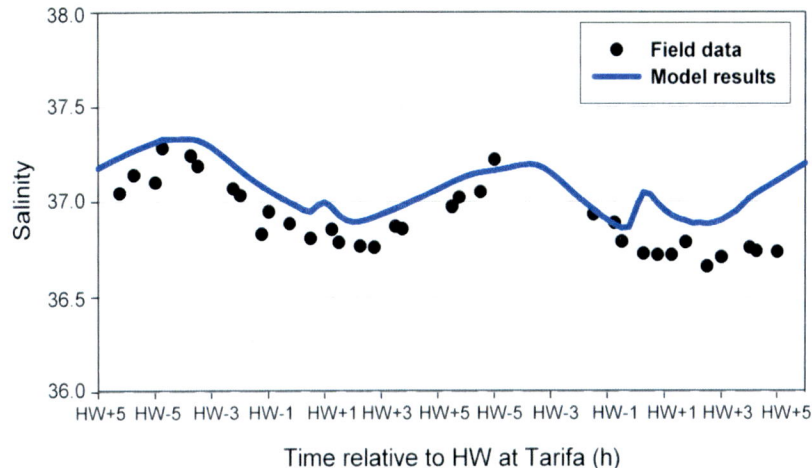

Figure 6. Measured (black dots) and modelled (blue line) upper layer salinity along 24 h. at the Eulerian station using the coupled model. Horizontal axis is time referred to the High Water (HW) at Tarifa.

This model could also be used to evaluate the temporal and spatial (in the along-strait direction) distribution of the interfacial mixing intensity (Figure 7). Unsurprisingly, there is intense and pulsating mixing over the Camarinal Sill although the most intense interfacial mixing occurs eastwards of Tarifa Narrows (see label in the bottom panel of Figure 7). This happens because in this part of the Strait the interface tends to be shallower (Bray et al, 1995) and thicker (García-Lafuente et al, 2002) than elsewhere. The upper layer becomes thinner gradually and, thereby, its velocity increases to satisfy mass conservation. Therefore, the velocity shear is enhanced in this region and the interfacial mixing increases. A similar result was suggested by Sannino et al (2004) when analysing water entrainment/detrainment forced by tides. As expected, two mixing-enhanced events happen every day, indicating the tidal periodicity of the phenomenon. They correspond to the increased shear that takes place during flood tide (García-Lafuente et al, 2000; Izquierdo et al, 2001; García-Lafuente et al, 2002).

However, there is some uncertainty in the estimates of the amount of interchange between layers driven by vertical mixing processes in the Strait of Gibraltar based on field measurements (Minas and Minas, 1993) due to the high variability of the physical environment. Estimates of nutrient fluxes given in the introduction for the case of no-mixing and steady exchange are not realistic due to the role that mixing plays in the inter-layer exchange of properties. Furthermore, the comparison exercise (between field data and model simulation) carried out for salinity demonstrates the necessity to include mixing in any model of the Strait in order to obtain realistic representation of the biogeochemical signatures in the area.

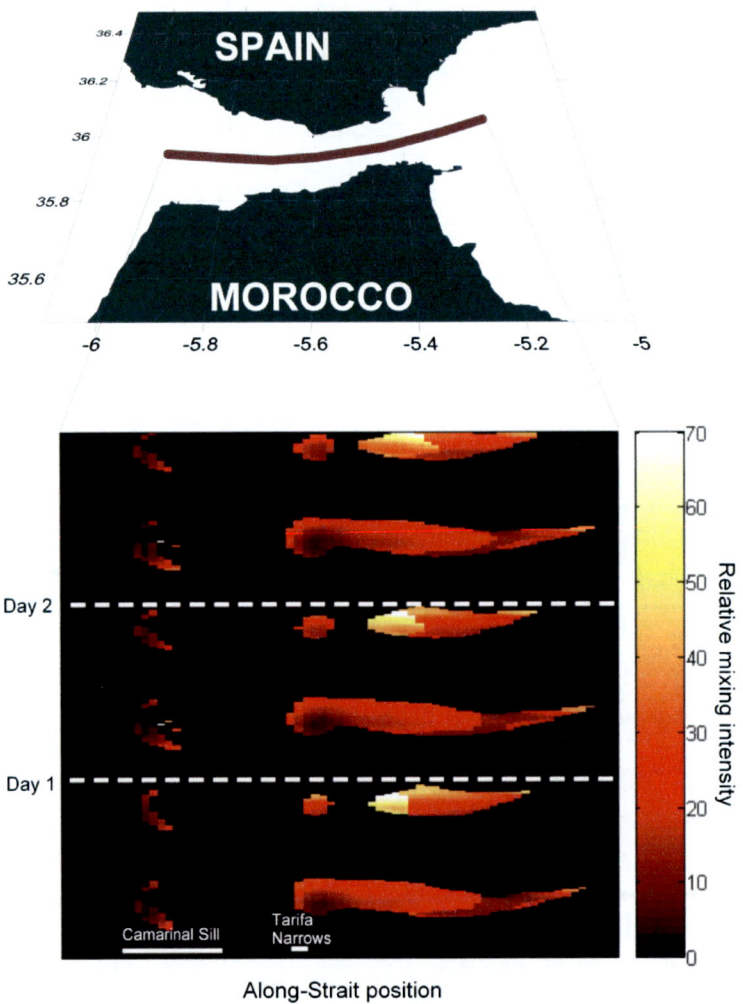

Figure 7. Mixing intensity (in relative units) between upper and intermediate layers during three days of simulations calculated according to the mixing-advection model.

Thereby, the coupled hydrodynamic-biogeochemical model could be used to estimate the amount of nitrogen in the outflowing layer that is introduced in the upper layer by mixing and hence, recirculates to the Mediterranean Sea (Figure 8). The concentration of nutrients in the upper layer of the model is a function of the tidal amplitude (figure 8b) and the percentage of the outflowing nitrate recirculating back into the Mediterranean due to interfacial mixing is significantly correlated with tidal amplitude ($r^2=0.7$; $p<0.01$), varying between 4% and 35% with an average of 16.3% (Macías et al., 2007).

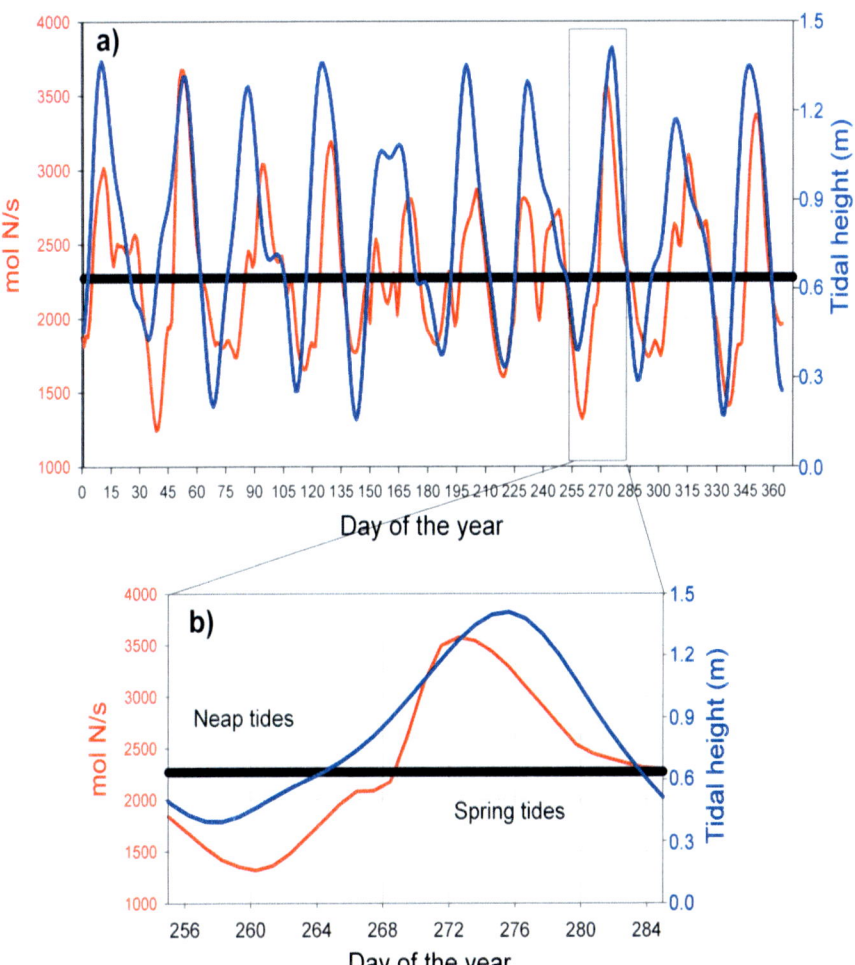

Figure 8. Mean daily flux of nutrient in the upper layer towards the Mediterranean (red line) and tidal height (blue line). a) During the whole simulation. b) During 30 days. Thick solid black line indicates the mean carbon nitrogen flux over all the period.

This mean value of recirculation is in good agreement with previous estimates based on field measurements (Wesson and Gregg, 1994; Gómez et al, 2000b; Dafner et al, 2003), although it is in the lower end of the measured range. One hypothesis for the slightly low value predicted by this model is its 1-D character, which neglects 2-D effects. Flow-topography interactions favour mixing and upwelling processes in the shallower marginal areas to the sides of the modelled transects. These processes, in turn inject nutrients laterally into the upper layer in a manner that the 1-D model is unable to reproduce.

Also, the only source of nutrients to the upper layer in the model is the mixing between the surface Atlantic layer and the deeper Mediterranean waters (see Macías et al, 2007). The contribution of the relatively nutrient-rich North Atlantic Central Water (Packard et al, 1988) is not considered, though it is known that this water could be upwelled during some phases of the tidal cycle, being incorporated into the main along strait circulation (e.g. Gómez et al, 2001; Macías et al, 2006).

These two processes not considered in the model will lead to surface nutrient concentration underestimation but the good agreement with the previously reported data based on field sampling indicates that the likely underestimation is not significant.

The year-averaged amount of nutrients introduced into the Mediterranean by the Atlantic inflow computed from the model is 2272.8 mol N/s (Figure 8), but it fluctuates between a minimum of around 1200 mol N/s and a maximum greater than 3500 mol N/s. The average value must be compared with the mean amount of nutrients advected into the Mediterranean Sea in the case of no-mixing, steady exchange as discussed in the introduction. The value of 972 mol N/s computed for that case increases to 2272.8 mol N/s if interfacial mixing is taken into account, that is, an increase of 130%. Obviously, mixing driven by tides cannot be neglected when computing nutrient fluxes. A consequence of this is that the estimates made with field data collected over a limited period of the year will differ from the real annual mean values by a factor dependent on the amplitude of the tide during the sampling period.

One of the most interesting effects of the tidally-induced mixing from a biological point of view is the fertilisation of illuminated surface waters. Tidally-induced mixing in the Strait supplies nutrients to the phytoplankton in the photic layer, which might be thought to result in a positive relation between tidal amplitude and phytoplankton abundance as described above for the field data (see Figure 5). However, a negative correlation ($r^2=0.2$; $p<0.01$) between tidal amplitude and phytoplankton concentration in the upper layer was predicted by the model, indicating that the phytoplankton is unable to utilise fully the amount of nutrient available through mixing.

An explanation for this behaviour is provided by the model simulations, which show that interfacial mixing produces strong dilution of the phytoplankton in the upper layer because of the low concentration of cells in the lower layer. The population growth (triggered by the nutrient injection) is not large enough to compensate losses due to mixing-related dilution and therefore the net concentration of phytoplankton in the upper layer decreases towards the east. The main reason of the low increase of phytoplankton concentration in the model along the Strait is the short residence time of cells within the model domain. An average phytoplankton cell only takes 25 hours to cross the Strait from west to east in the upper layer and, as the maximum growth rate of the modelled phytoplankton is 3.0 d^{-1} and, considering the small initial concentrations of phytoplankton, there is no time for sufficient growth of the population regardless of the amount of nutrients available.

When comparing the temporal dynamics of measured and modelled phytoplankton concentration at the eastern side of the Strait (Figure 9) appreciable differences can be observed. In this figure it is clear that observations exhibit two peaks of high concentration around HW-4 simultaneous to the peaks of salinity shown in Figure 6, although the distribution of phytoplankton has higher dispersion than that of salinity. The peaks are twelve hours apart, indicating tidal periodicity in the forcing of the phytoplankton peaks. Although the model output shows similar behaviour, it has two important drawbacks. First, it must be noticed that scales for modelled (left) and observed (right) concentrations in Figure 9 are different, so the observed concentration of phytoplankton is about twice larger than the concentration predicted by the model. Second, there is also a clear difference in the time when predicted and observed maxima appear, the maximum observed concentration coinciding with a minimum of predicted abundance and vice versa (correlation coefficient $r^2 = 0.3$, $p<0.01$). So, although the model predicts the spatial distribution of salinity reasonably well, is not able to predict phytoplankton abundance satisfactorily (more details about this discussion can be obtained in Macías et al, 2007).

With the residence times predicted by the model, it is clear that the high levels of chlorophyll measured in the eastern section of the Strait must be the result of something other than local growth. Another source of chlorophyll is necessary in order to explain the high concentration observed in the eastern section.

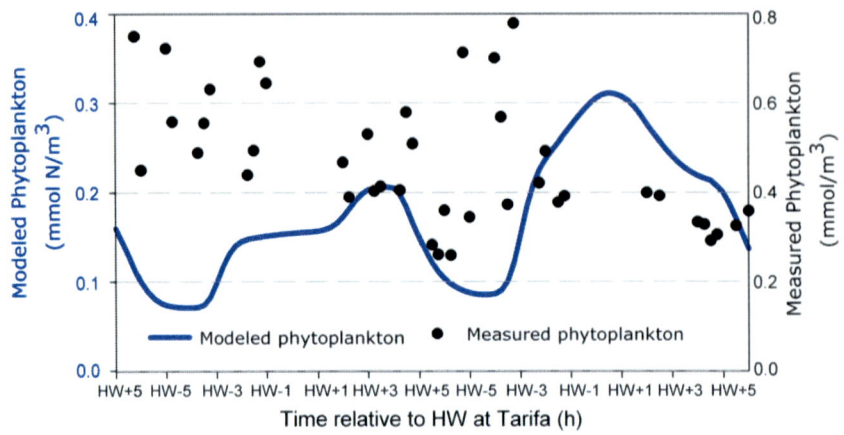

Figure 9. Measured (black dots) and modelled (blue line) Phytoplankton concentration (mmol N/m^3) in the upper layer along 24 h. at the Eulerian station.

3.2.3. Some Conclusions About the 1D Dynamics

From the comparison between field data and modelling results shown above some conclusions emerge. On the one hand, the 1D simplification of the Strait seems to be good enough to represent the general dynamics of some tracers such as salinity or nutrients as both the temporal evolution and absolute values coincides with the reported values from field sampling.

However, important and large differences between observation and predictions were found when phytoplankton concentrations are compared. This seems to be incongruent with the previous paragraph and could lead to suspect a possible misrepresentation of the biogeochemical compartment by the simplified NPZ model used in the simulations. However, changing the complexity of the biogeochemical model (see Macias et al, 2007) does not lead to an improvement of the fitting between observations and predictions.

The high chlorophyll concentration in the eastern side of the Strait is recurrently shown by field data (e.g. Minas et al, 1991; Gómez et al, 2001; Echevarría et al, 2002; Macias et al., 2006). Local growth of phytoplankton populations while crossing the Strait in an environment enriched by the mixing-driven injection of nutrients in Camarinal Sill has previously been put forward to explain these observations (Reul et al, 2002).

Other works (Packard et al, 1988; Ruiz et al, 2001) propose that noticeable increase of phytoplankton concentrations in the Atlantic layer away from the

Strait can be better explained by the influence of mixing in the Strait than by local enrichment in the Alboran basin, due to the usual large advection velocities of the Atlantic jet that would imply the observation of these effects in the north-western Alborán Sea or, even, much further west, in the Almería-Orán front, at the east of the Alborán Sea (Arnone et al, 1990). These latter results would agree with the predictions of the model, which indicates a residence time in the main channel too short to allow for a significant growth of phytoplankton during its transit through the Strait, being much more probable to register noticeable results of growth within the Alborán basin itself.

An alternative hypothesis to explain the lack of concordance between observed and modelled phytoplankton in the eastern section of the Strait is related with tidally-induced dynamics of the surface layer in the western part of the Strait. In this region the currents in the surface layer reverse during part of the semidiurnal tidal cycle (Béthoux and Copin-Montégut, 1986; Candela, 1990) while they hardly ever reverse east of Tarifa Narrows (García-Lafuente et al, 2000), giving rise to intense divergences in the upper layer. The physical model used to force the biogeochemistry reproduces this phenomenon and indicates that the divergence takes place between Tarifa Narrows and Camarinal sill around high water time (Figure 10).

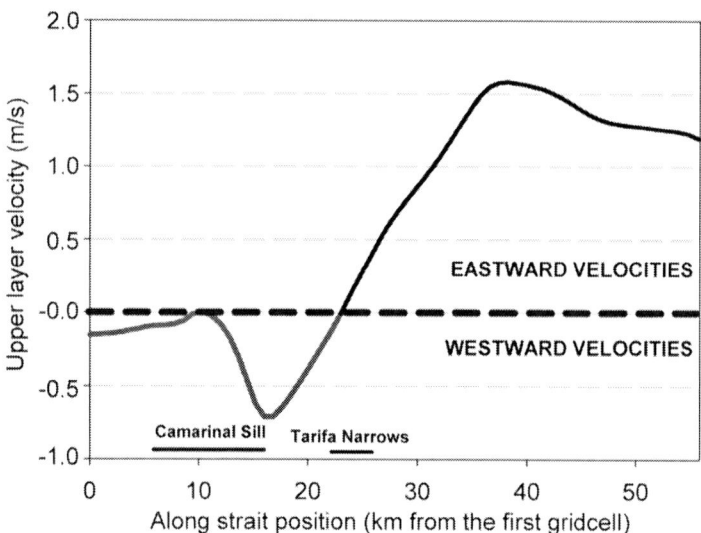

Figure 10. Upper layer velocity (m/s) along the central section near the High Water time.

Internal oscillations of the interface can partly account for the upper layer divergence (García-Lafuente et al, 2000) but some compensation by horizontal flow from both north and south coastal areas cannot be disregarded. The input from the chlorophyll-rich coastal surface waters into the central channel would increase the chlorophyll concentration periodically by means of a mechanism not specified in the used model. Some evidence of suction of coastal waters through the Strait is provided by Van Geen et al (1988), who suggest that the high concentrations of trace metals in the Mediterranean surface waters when compared with open ocean Atlantic waters could be explained by the advection through the Strait of Gibraltar of waters from the continental shelf of the Gulf of Cadiz, which are mainly enriched by the Tinto and Odiel rivers discharges (Elbaz-Poulichet et al, 2001).

Horizontal advection from the north and south is also supported by the following discussion. If there is no horizontal advection, salinity and phytoplankton concentration must be negatively correlated because salinity increases when interfacial mixing is enhanced but phytoplankton concentration diminishes by dilution. The model predicts this behaviour but observations behave in the opposite manner. The contradiction can be overcome if simultaneously with intense interfacial mixing there is advection of coastal chlorophyll-rich waters to the interior of the channel. Since interfacial mixing is favoured by the undulatory processes in Camarinal (Wesson and Gregg, 1994) and divergence usually happens when undulatory processes are active (Izquierdo et al, 2001), it follows that interfacial mixing and divergences are simultaneous and, hence, horizontal advection and intense mixing occurs at the same time. Notice that horizontal advection has no impact on salinity since coastal and open channel waters have similar values (Navarro, 2004; García-Lafuente and Ruiz, 2007).

The coincidence of the highest concentrations of chlorophyll with the presence of undulations would lead to a packaging of the cells in patches as such undulations are able to create convergence and divergence alternating areas. Those concentration mechanisms could, thereby, explain the high-chlorophyll patches observed in the diel cycles carried out in the eastern sector of the channel of the Strait.

The mechanisms, hypotheses and proofs related with this 2D horizontal interaction between the coastal areas and the main channel of the Strait will be further discussed in the last chapter of the book.

3.3. 2D-TIDALLY-INDUCED VERTICAL PATTERN OF WATER MASSES CIRCULATION AND BIOGEOCHEMICAL SIGNATURES

What is presented above in this chapter is but a simplification of the biogeochemical signatures associated with the tidal dynamics in the Strait of Gibraltar as this physical-biological coupling should also have consequences on the vertical distribution of the planktonic biomass. In this sub-section we will focus on describing the different types of planktonic biomass accumulation that can be observed in the Strait as well as its tidally-forced dynamics while travelling between basins following the main results presented by Macías et al. (2008).

3.3.1. Identified Chlorophyll Maxima

From the observations performed along the main channel of the Strait (see Figure 3a) at least three different types of chlorophyll maxima could be identified, each with different absolute chlorophyll values, position within the water column and physiological properties. The three types of maxima were: (i) those associated with the SAW-NACW interface, (ii) those situated between SAW and MOW, and (iii) those found in the transition between NACW and MOW (Macías et al., 2008). The mean characteristics of each type of maximum are summarised in Figure 11, where the confidence intervals calculated using the t-student analysis (Zar, 1984) are also shown. There was a clear relationship between chlorophyll and carbon concentration for all the identified maxima, with correlation coefficients of > 0.9 ($p < 0.001$) in all cases. Thus, photoadaptation did not seem to be a key factor determining the presence of the DCM suggesting that the occurrence of the maxima could be attributable to an increase in phytoplankton biomass due to accumulation of cells at the water interfaces.

The interface between SAW and NACW is usually characterised by a strong thermocline between the surface and the central waters (Gascard and Richez, 1985; Ochoa and Bray, 1991). A chlorophyll maximum linked to that shallow thermocline has been previously registered in the western section of the Strait (Gómez et al, 2000; Echevarría et al, 2002) and in the Gulf of Cadiz (Navarro et al, 2006). Because of this, the denomination of 'Atlantic maxima' (**AM**) was used to define this SAW-NACW chlorophyll maximum.

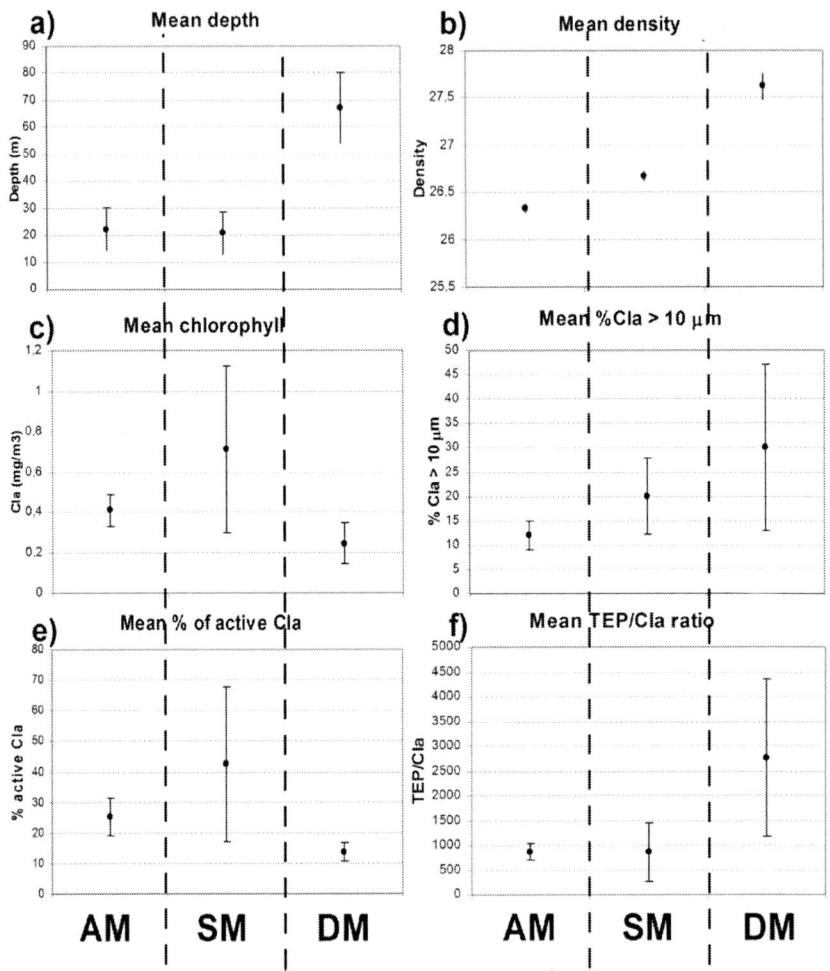

Figure 11. Mean characteristics and 95% confidence intervals (t-student analysis) of the identified maxima.

Navarro et al (2006) suggested that the chlorophyll maxima associated with that interface was attributed to a functional response of the picoplankton to the existence of a nutricline between the nutrient-rich NACW and the impoverished SAW (Herbland and Voituriez, 1979). This hypothesis was supported by the observations made within the Strait as mean Nitrate + Nitrite (NN) concentration in the overlying waters was 0.6 μM ($n = 26$; CV = 70 %), a concentration below the threshold of 1 μM accepted as limiting for phytoplankton proliferation (Eppley et al, 1969; Wash, 1988), whereas NN concentration in the deeper water layer (between the 26.35 and 26.5 isopicnals) presented a mean value of 2.0 µM ($n = 21$; CV = 76 %), confirming the existence of a marked nutricline in this region of the water column.

The predominance of small cells in the phytoplankton assemblage of the AM (Figure 11) is typical of oligotrophic environments (Li, 2002) were maxima are usually associated with gradients in nutrients distribution (Moran et al, 2001). Picoplankton composition also support this hypothesis, as *Prochlorococcus* (the more abundant picoplankter in AM) has been reported to be more abundant in offshore waters (Goericke et al, 2000) and is prevalent in the open region of the Gulf of Cadiz (Reul et al, 2006; Echevarría et al, 2009). The low percentage of active chlorophyll and the high TEP/Chla ratio (Figure 11) of these maxima are indicative of a senescent phytoplankton population; mucilage production is known to be a good indicator of declining plankton communities (Prieto et al, 2006). This extent is confirmed by production measurements using C^{14} incorporation ratios by Macias et al (2009) as a low maximum production (Pm) and initial slope of the production-irradiance curve (α) was found in those AM.

The larger chlorophyll maxima (Figure 11) were those found associated with the SAW-MOW interface being classified as "Suction Maxima" (**SM**). These maxima correspond with the previously observed in the diel cycle sampling (see above) and do not seem to be linked to the presence of a nutricline, as NN concentration in the overlaying waters was around 2.2 µM ($n = 9$, CV = 53 %), well above the threshold limiting value that limits phytoplankton growth. Therefore, the generation of these chlorophyll patches is likely to be related to physical processes, rather than *in-situ* growth.

The higher mean concentration of chlorophyll, percentage of larger cells and active chlorophyll in these maxima give further support to the coastal origin' hypothesis presented above, as larger cells are generally found in coastal environments where nutrient inputs to the surface layer are more frequent (Malone, 1980). Moreover, the increase in the quantity of *Synechococcus* and the decrease of *Prochlorococcus* (see Macías et al, 2008) also supports the coastal origin of these chlorophyll patches as *Synechococcus* occurs mainly in shallow inshore areas (Chisholm, 1992; Goericke et al, 2000; Pan et al., 2004) and it is usually found in the coastal region of the Gulf of Cadiz (Huertas et al, 2005; Reul et al, 2006; Echevarría et al, 2009). Also, the relative low TEP/Chla ratio in SM is an indicator of a population actively growing as it has been documented that in early bloom stages the production of TEP is minor (Corzo et al, 2000). These SM also presented higher Pm and α in the C^{14} incubation experiments performed by Macías et al (2009).

The origin of the last chlorophyll maximum (the Deep Maxima (**DM**)) is not clear, but examination of their characteristics suggests that they are generated by sedimentation of large aggregates (high percentage of chlorophyll in particles larger than 10 μm) formed by senescent cells (low quantity of active chlorophyll, high TEP/chla ratio and extremely low values of Pm and α in the C^{14} incorporation measurements). These accumulate at the interface due to a reduction in their settling velocity caused by either the density gradient created by the strong halocline (Longhurst and Harrison, 1989; Margalef, 1989) or because of modifications in turbulence levels (Ruiz et al, 2004). The accumulation of larger particles at this interface, observed using an Underwater Video Profiler (UVP), has been previously described by Gómez et al (2001).

3.3.2. Tidal Dependence of Water Masses Circulation and DCM Positions

As circulation patterns of water masses through the Strait are expected to influence the position and, even, the composition of the chlorophyll maxima, the conceptual diagrams of along-strait water masses and DCM position during different phases of the tidal cycle created by Macías et al (2008) are presented in Figure 12. Vertical scale maintains the original values recorded in the CTD casts (position and thickness of the interfaces and DCM) while horizontal scale has been modified to include the entire longitude of the Strait.

As already pointed out, shortly before the High Water time (Figure 12, upper plate) there is an enhanced outflow (westwards) of deeper water over the Camarinal Sill towards Atlantic (García-Lafuente and Vargas, 2003). When the tidal amplitude is high enough, even the upper layer can flow towards the west (LaViolette and Arnone, 1988; Candela, 1990) creating a surface divergence zone over the Camarinal Sill (see Figure 10), which is thought to be responsible for the entrainment of coastal waters into the central region of the Strait. At this stage of the tidal cycle internal waves have already formed (if the tidal amplitude is sufficient), but remain arrested over the Camarinal Sill because propagation against the main outflowing flow is impeded.

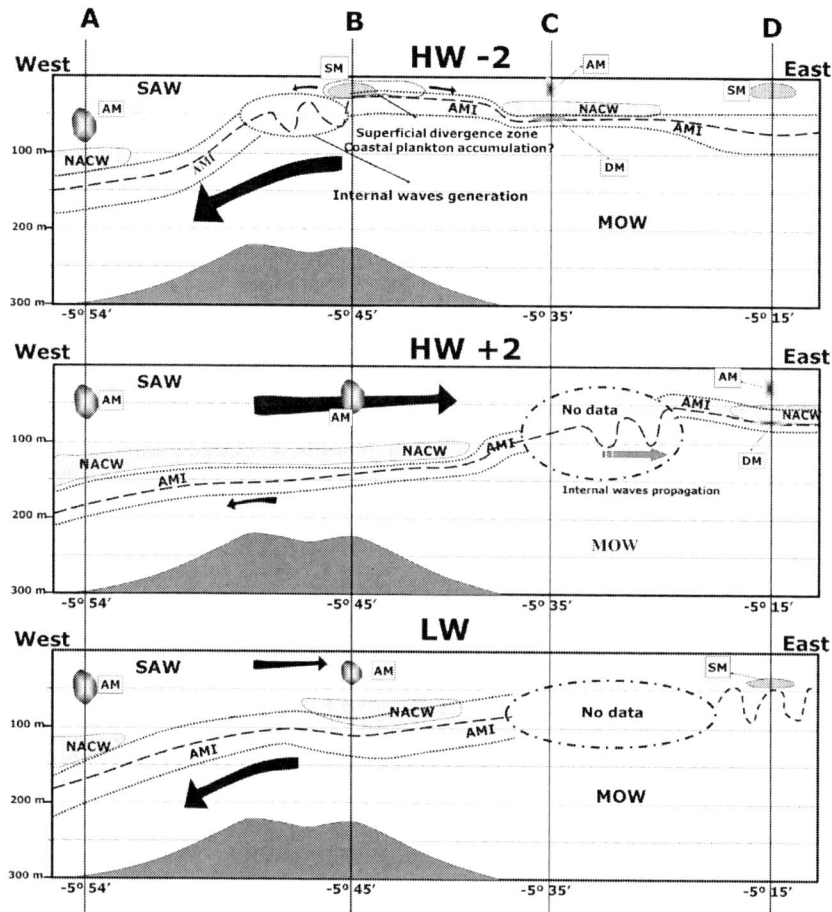

Figure 12. Conceptual diagram showing interface and DCM positions and water mass circulation within the Strait during different stages of the tidal cycle. Areas encircled by a dashed-dotted line represent values obtained from the literature (no data available).

West of the Sill (position **A,** Figure 12) SAW is situated in the upper 100 m of the water column, followed by NACW between 100 and 150 m and MOW at depths of < 150 m. Maximum chlorophyll is observed at the SAW–NACW interface situated at 75 m and corresponds to the AM. During this tidal stage, only SAW and MOW are detected over the Camarinal Sill (position B, Figure 12), possibly due to the current regime, which prevents NACW entering this area. Surface chlorophyll maximum observed in this area are associated with the SAW-MOW interface, with the characteristics of the SM (i.e. with a coastal origin). Towards the central region of the Strait (position C, Figure 12), the three water masses are again detected in the water column. In this situation, NACW forms a "tongue" lying over MOW and below SAW, which is located only in the central area of the Strait; no NACW signal is present at positions "B" or "D" (Figure 12). The NACW tongue found at intermediate depths at station "C" during this tidal stage, was originated in the previous tidal cycle when, in a situation similar to that shown in the lower plate of Figure 12 (LW time), the NACW is able to overcome the sill after the flow eastward is re-established being dragged by the main surface current towards the Mediterranean side. At this position "C" two chlorophyll maxima are commonly observed, an AM at the upper interface and a DM at the lower. On the eastern side of the Strait (position D, Figure 12), only SAW and MOW are commonly observed. In this area, the chlorophyll maximum linked with the interface appears to be very similar to those found over the Sill (position B), i.e., a coastally-originated SM.

Four hours later (i.e two hours after the high water time) the inflow of Atlantic water is reinforced and a weakening of the Mediterranean outflow takes place (Figure 12, intermediate plate) (Armi and Farmer, 1988). The internal waves have been already released during HW and therefore, the NACW can cross the sill and flow towards the Mediterranean Sea.

West of the Camarinal Sill (position A, Figure 12), the three water masses (SAW, NACW and MOW) are detected and an AM is observed at the SAW-NACW interface. Above the Camarinal Sill (position B) the three water masses are still present, a clear NACW signal is observed between 120 and 130 meters and an AM is still associated with the SAW-NACW interface. During this stage of the tidal cycle the internal waves are propagating within the Strait, and are expected to be around position C, probably with the SM described in position "B" of Figure 12, upper plate (HW-2). However, there was no data available for this section of the channel during this tidal stage to confirm this assumption. At position "D", the tongue of NACW described in region "C" of Figure 12, upper plate was also detected. Furthermore, the two maxima linked to the interfaces SAW-NACW and NACW-MOW were again clearly observed in the CTD profile.

Finally, four hours later coinciding with the low water time the internal waves are expected to be close to the eastern side of the Strait (Farmer and Armi, 1986; Macías et al, 2006). At this stage, the current velocity of the surface layer over the Sill progressively decreases and the deeper outflow is reinforced. In this situation a shallowing of the AMI takes place, obstructing the eastward flow of NACW (Figure 12, lower plate).

In position "A" (west of the Sill), the composition of the water column is very similar to that described in the two previous diagrams i.e., the presence of all three water masses and an AM at the upper interface. The signature of NACW was still detectable in the profiles performed over the Camarinal Sill (position B). Four hours later, NACW will form the tongue described in position "C" at HW-2 (Figure 12, upper plate). An AM was registered between the SAW and the NACW. During this phase of the tidal cycle the internal waves generated in the previous outflowing event are expected to be approaching the eastern side of the Strait (position D). In the cast performed at this position there was no NACW signal and a shallow SM connected to the SAW-MOW interface was observed. This maximum was very similar to the DCM associated with the generation of internal waves located over the Camarinal Sill at HW-2 (position B, Figure 12, upper plate). Completing the tidal cycle and four hours after the situation above, the scenario was again comparable to the one described in the upper plate of Figure 12.

These results clearly indicated that the nature of the DCM observed in the Strait of Gibraltar is strongly influenced by tidal amplitude, so a discontinuous entrance of different phytoplankton assemblages into the Alboran Sea would be expected (Macías et al, 2008). This discontinuous entrance of biogeochemical material through the Strait is important for the pelagic ecosystem of the Alboran Sea, as a pulsating input of both plankton and nutrients could drive the behaviour of the system. This extent has been recently addressed by Macías et al. (*2010a*) using numerical models to explore how changing from a continuous to a pulsating nutrient input (while maintaining constant the total amount supplied) affects the dynamics and structure of the pelagic community. These last authors showed that a discontinuous pattern of nutrient input to the ecosystem similar to the one happening at the Strait induce larger biomass production in the system and also differentially favours some plankton groups. For example, under fast and repetitive nutrient pulses (like the ones associated with the tidal mixing at the Strait) favours fast-growing and big cells when compared to the continuous supply case. This has consequences on trophic web dynamics as energy flow would be more or less efficient depending on the biomass distribution between groups (Smetacek, 1999) and, thereby, should have consequences on the higher trophic levels of the Alboran basin. So, the nature of this input should be taken into account when explaining the spatial and temporal patterns of biological constituents of the Alboran ecosystem.

In summary, tidal induced variability within the Strait of Gibraltar not only affects the transport and circulation of the water masses that meet in the area, but also the distribution and characteristics of chlorophyll patches. This fact corroborates the importance of tidal induced phenomena in explaining physical and biological structures at the regional scale.

Chapter 4

HYPOTHESES, EVIDENCES AND PROOFS OF THE TIDALLY-INDUCED COASTAL-CHANNEL INTERACTIONS IN THE STRAIT

Along the entire book it has been stressed that the hydrodynamic and associated biogeochemical pattern within the Strait of Gibraltar is mainly controlled by the tidal dynamics through various and complicated processes including changing advection directions, vertical interfacial mixing and (probably) complex coastal-channel interactions.

This last process is particularly relevant for the pelagic ecosystem of the adjacent Alboran Sea as it implies the injection of coastal elements into the open sea at the same time that the surface layer (i.e. the lighted one) is being fertilized by the interfacial mixing associated to the internal wave's generation. The coincidence of high nutrient concentration with high plankton biomass should affect the structure and dynamics of the entire pelagic ecosystem not only within the Strait itself but also on the nearby North-Western Alboran Sea.

This mechanism of coastal-channel interaction had been unnoticed until the performance of several exhaustive samplings at the beginning of the last decade, some of them already commented in the previous sections of the book. In this last chapter we will focus on direct evidences and proofs of the exact mechanism of such interactions based on both field data samplings, tracers release experiences and modelling simulations.

In the first part of this last chapter we will focus in describing some results on how plankton distribution and community composition is affected by the generation and development of the surface divergence and internal waves based on observations made on the Camarinal Sill area and presented in Macías et al (2010b).

The second part will be a description of an *ad-hoc* experiment designed to confirm the existence of these interactions by combining a water-tracer release experience with computing simulation using one of the latest most advanced 3D hydrodynamic models of the Strait available nowadays.

4.1. PLANKTON DISTRIBUTION PATTERN AND PHYSICAL FORCING AT THE CAMARINAL SILL

During a cruise carried out in November 2003 on board BIO Hespérides, a specific zooplankton sampling was designed over the Camarinal Sill trying to study the effect of the internal wave's generation and development on the planktonic community distribution and composition (see details in Macías et al, 2010b).

Zooplankton was sampled using a Longhurst-Hardy Plankton Recorder (LHPR) net, which is a multiple sampling net designed for plankton capture taking several samples at each cast (Wiebe and Benfield, 2003). Collected samples were analysed using a software for plankton determination known as PVA (Plankton Virtual Analyzer) which counts and measures planktonic organisms (Boyra et al, 2005).

The LHPR was also equipped with a CTD SBE-19 and fluorometer sensor that registered pressure, temperature, salinity and fluorescence of the water masses as the LHPR passed. This design allows a large number of samples to be obtained in a single haul (between 30and 60), and is able to resolve partially the variability of zooplankton distribution at short scale and the aggregation phenomena that were expected to occur in the Strait.

Two casts were performed over the Camarinal Sill (Figure 13) just before (cast 1) and after (cast 2) the generation of internal waves (Figure 14). During cast 1 the tidal current changes from eastward to westward and the cast finished shortly after the low-water time. Tidal currents were towards the west during the entire period of cast 2, since it coincided with the tidal phase when the internal waves are being developed over the Camarinal Sill. This cast ended just before the high-water time, when these waves are typically released (Figure 14).

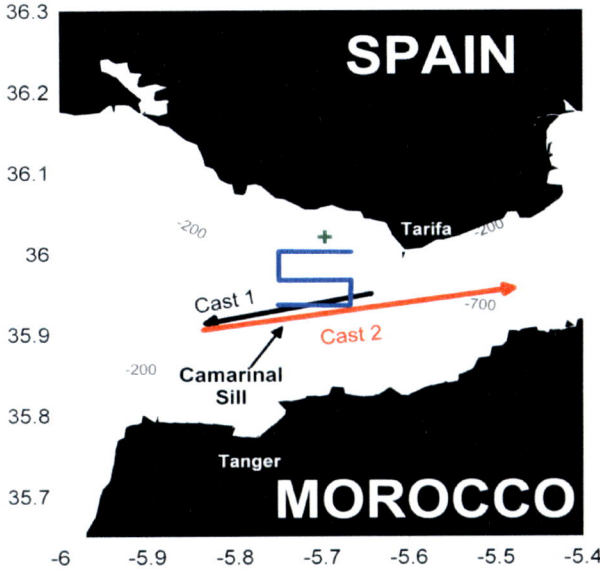

Figure 13. Position of the Rhodamine initial release (green cross). Path followed by the vessel during the Rhodamine sampling (blue line). Westward (black line) and eastward (red line) casts of the LHPR sampling over the Camarinal Sill.

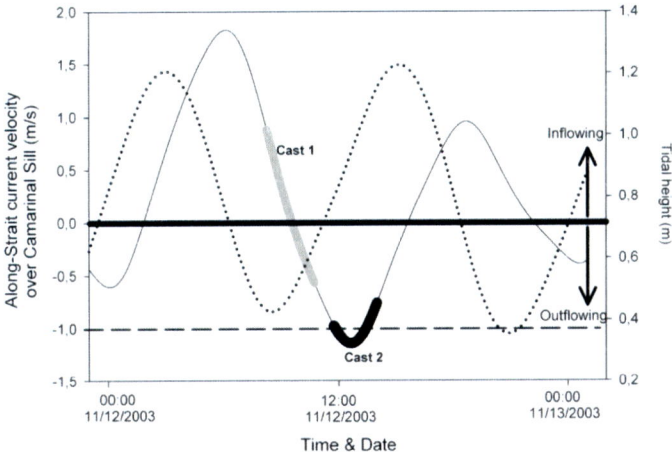

Figure 14. Tidal conditions over the Camarinal Sill during the LHPR sampling period. Solid line represents current velocity over the sill (positive are westward). Dotted line marks the tidal height at Tarifa harbor. Dashed line indicates minimum outflowing velocity for the generation and liberation of internal waves (according to Vazquez et al., 2008).

From the collected samples total plankton biomass was determined by using a wet weight measurement while community composition was assessed by using the PVA software and a discriminant analysis based on the Fisher's functions (Krzanowski, 1998).

4.1.1. Cast 1 (Westward Hauling- Eastward Tidal Flow)

This entire cast has been divided into five sectors (designated A to E in Figure 15) based on the spatio-temporal distribution of temperature, salinity and chlorophyll.

In sector **A** (east of the sill), NACW was located around 100 meters depth marked by a salinity minimum in Figure 15. The transition between sectors A and B is indicated by a strong horizontal salinity gradient (Figure 15), and a loss of the NACW signal. Sector **B** is characterized by a temperature gradient from surface to deeper waters while salinity is very constant from the surface down to 100 meters. Between sectors B and **C**, a new horizontal gradient in salinity values is present (Figure 15) with increasing values towards the west. The location of sector **D** was directly above the Camarinal Sill, and presents a marked horizontal salinity gradient (Figure 15) increasing from 36.5 up to 37.5, and no vertical structure. Sector **E** starts with a sharp change in salinity conditions of the upper water layer, from a uniform value of 37.5 to a steep vertical gradient (Figure 15). A salinity minimum of 35.5 is observed at 50 meters depth, indicating the presence of NACW and absence of MOW.

Chlorophyll distribution (Figure 15, lower plate) shows a DCM located around 50 meters depth and extending from sectors C to E. In general chlorophyll concentration is relatively low (average 0.9 mg m^{-3}) and even the DCM is not very marked, with values of 1 - 1.3 mg m^{-3}.

Zooplankton biomass along this cast does not seem to be correlated with any of the hydrological (temperature or salinity) or biogeochemical (chlorophyll) variables (Figure 16, upper plate) being the average biomass particularly low for all this cast. Composition of the zooplankton community during this sampling was analysed in 36 samples taken in sectors B to E. Very little differences are found between samples, indicating a very similar composition of the zooplankton community throughout the sampling path.

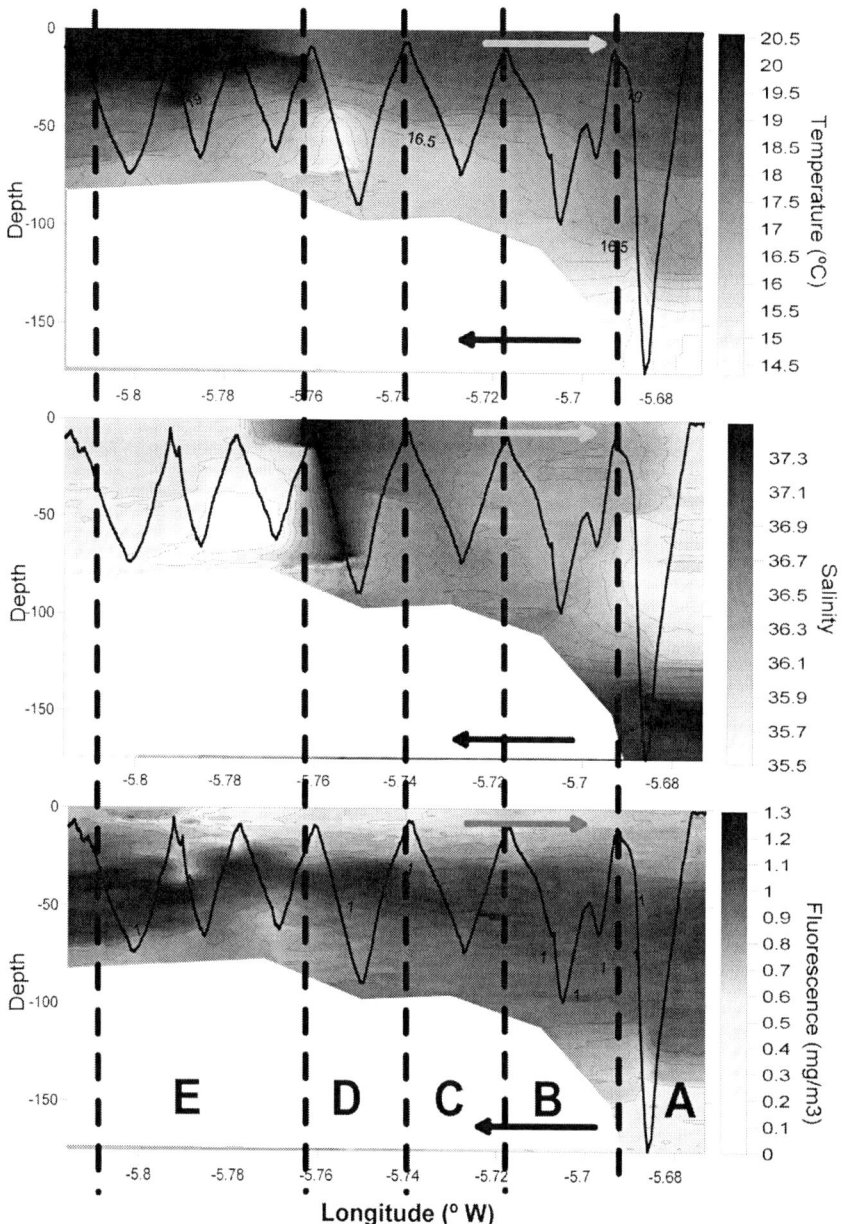

Figure 15. Temperature (°C), salinity and chlorophyll (mg/m^3) distributions during cast 1. Grey arrows show water current direction and black arrows show sampling direction. In sector E water flow velocity decreased and even changed its direction (i.e. flowing eastward) as indicated in figure 14.

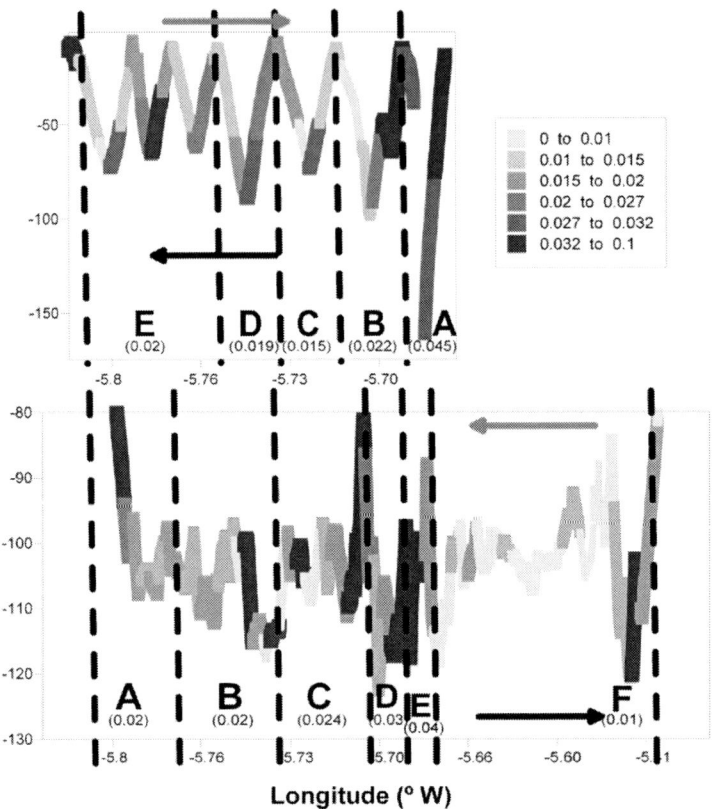

Figure 16. Zooplankton biomass (g/m^3) during cast 1 (upper plate) and cast 2 (lower plate). Values in parenthesis are mean biomass concentration in each sector. Grey arrows show water current direction and black arrows show sampling direction.

4.1.2. Cast 2 (Eastward Hauling- Westward Tidal Flow)

This second cast was performed entirely during the west-flowing phase of the tidal cycle (Figure 14) and a different strategy was used: the LHPR was hauled from east to west (Figure 13) at a constant depth (nominal depth 100 m).

As in the previous casts, it has been divided into 6 sectors taking into account the hydrological conditions of the water column (mainly water masses, Figure 17). Sector **A**, situated west of the Camarinal Sill is characterised by a constant temperature and salinity (Figure 17) with the presence of NACW and low values of chlorophyll throughout the whole sector

(Figure 17, lower plate). Further east, in sector **B**, a gradient of both temperature and salinity is recorded with an increase in salinity and a decrease in temperature towards the eastern sectors (Figure 17).The NACW signal disappears, and even lower chlorophyll values are found (Figure 17, lower plate).

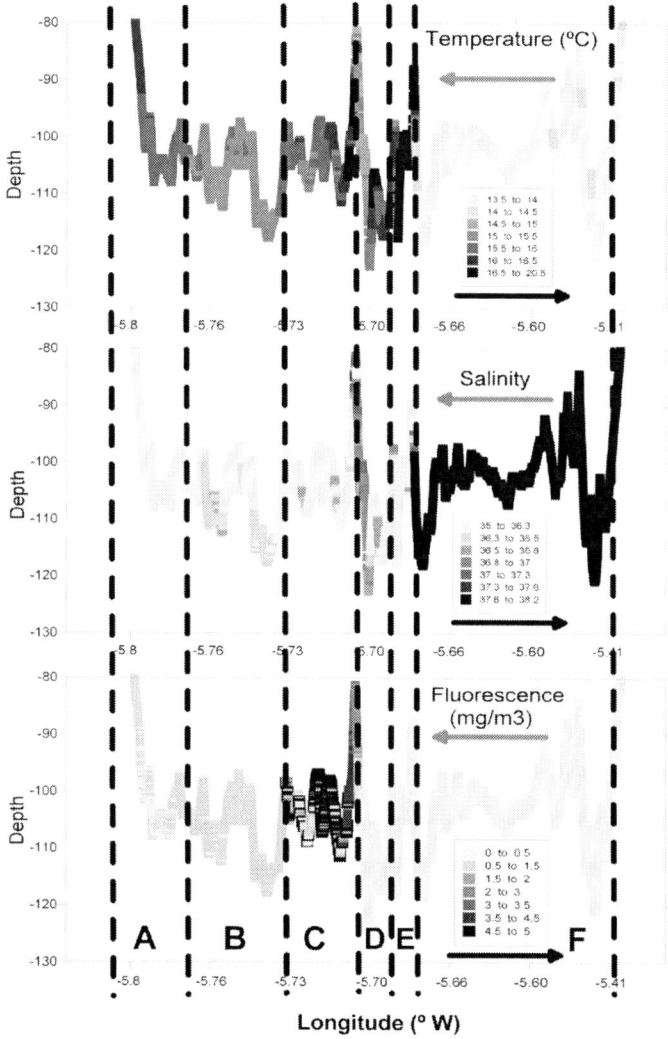

Figure 17. Temperature (°C), salinity and chlorophyll (mg/m^3) distributions during cast 2. Grey arrows show water current direction and black arrows show sampling direction.

The boundary between sectors B and **C** is located directly over the Camarinal Sill and abrupt changes in the depth of the net occur in this transition zone, from 120 m at the end of sector B and just 100 meters in the beginning of sector C (Figure 17). An approximate vertical velocity of around 0.3 m/s can be calculated from this vertical displacement of the net. There is a slight increase in salinity values and a decrease in temperature with respect to the preceding sector (Figure 17). The most dramatic change corresponds to the chlorophyll concentration, which is 10 times greater than in the preceding sectors (Figure 17). The chlorophyll concentration recorded in sector C is the highest found during all the casts carried out in the Strait.

Between sectors C and **D** strong vertical oscillations of the LHPR net are again observed (Figure 17), changing from 80 to 120 meters very rapidly, with associated vertical velocities of around 0.4 m/s. Corresponding with this oscillation an abrupt change to colder and more saline waters is recorded (Figure 17). At the same time, chlorophyll concentration decreases very rapidly (Figure 17). At the beginning of sector **E** the LHPR again undergoes sharp vertical movements and salinity values decrease whereas temperature increases (Figure 17). As for the preceding sector, chlorophyll concentration remains very close to zero (Figure 17). Finally, in sector **F** (which starts also with a severe oscillation of the LHPR) water composition changes. The salinity increases and the temperature decreases (Figure 17), indicating that only MOW is present in this zone. No changes in chlorophyll concentration are detected, and the values remain very low.

During this cast, zooplankton biomass tends to accumulate in the transition zones between the different sectors defined previously (Figure 16, lower plate).The density is greater in those transition zones characterised by marked horizontal gradients in salinity and/or temperature (such as the C-D and D-E boundaries). Taxonomic composition of the community during this last cast shows considerable differences in the various sectors, and is unlike the composition found in the previous sampling. Although copepods are still the most important group in the smaller fraction (<1000 µm), there are a number of different taxa showing increased abundance, such as ostracods, decapods larvae, etc. Within this smaller size fraction the second most abundant group (ostracods) show big differences, accounting for 20% in sectors A and F (the first and last sampled) but decreasing to less than 2% in sector E. In the vicinity of the Camarinal Sill (sections C and D), groups like siphonophores, salps and larvaceans (appendicularia) are more abundant than in the rest of the cast, in both the larger and smaller fractions. Finally, in sectors C, D and E mucilaginous material was present in the samples.

4.1.3. Evidences of Quick Changes and Coastal-Channel Interactions over the Camarinal Sill

For the first time the most dynamic area of the entire Strait (the Camarinal Sill sector) was sampled at two very different phases of the tidal cycle and internal wave development (Macías et al, 2010b). So it makes it possible to relate the hydrological environment to the biogeochemical signatures recorded during each cast and confirms the hypothesis of coastal-channel interactions forced by the tidal dynamics.

Phytoplankton distribution during the first cast over the sill could be analysed within the framework of a model of currents-plankton interactions developed by Franks (1992) for frontal areas. Following this model, when a divergent (horizontally) flux is present in a frontal zone the plankton tends to show accumulation in a horizontal band that crosses the pycnoclines. In this case it has been postulated that a surface divergent area could appear between the Camarinal Sill and the Tarifa Narrows (see Figure 13) coinciding with the tidal reversion (Izquierdo et al, 2001, Macías et al, 2007).

Predictions of the Franks' (1992) model agree quite well with the phytoplankton distribution found during this cast, since a high chlorophyll band was situated at about 50 meters depth from sectors B to D (Figure 15, lower plate). At the same time, the average concentration during the entire cast is quite low, corresponding to a situation of divergence. This same hydrological condition could be responsible for the generalised decrease in the zooplankton biomass recorded during this cast (Figure 16, upper panel). As these organisms possess swimming capabilities, the Franks' (1992) model predicts an effect different from the one observed for phytoplankton. In this case, a progressive reduction of the population density from the point of divergence is expected, with the faster swimmers located further from that point. This is observed in our data, as mean zooplankton concentration is very low over the Sill (sector C) but is higher to the west and east (Figure 16a). In almost all the samples from this cast the dominant organisms are copepods of the genus *Oithona* which has a relatively slow swimming speed (0.4 mm s^{-1}, Yamasaki and Squires, 1996). Larger and faster swimmers, like chaetognaths (swimming speeds of 1-2 cm s^{-1}, Mutlu, 2006), big euphausiids and siphonophores (30 cm s^{-1}; Bone, 2005) were scarce and only found in the samples from sectors B and E (the furthest from the area of divergence).

On the other hand, hydrological conditions in the Camarinal Sill area were substantially different during the second sampling (see Figure 14). In such tidal conditions and with enough tidal amplitude (tidal current over the Sill above -1m/s) internal waves are created along the AMI by the interaction of tidal currents and sill topography (Armi and Farmer, 1988, Bruno et al, 2002). Such undulatory processes remain arrested over the Sill until the outflowing velocity falls below -0.5 m/s (Vázquez et al, 2008) which happens at the end of this sampling (Figure 14).

The presence of the internal waves throughout this entire cast is revealed by the oscillation of the LHPR at the sector boundaries (Figure 17), since the sampling strategy in this cast was to maintain the net at a constant depth of around 100 meters. A schematic representation of the crest and trough of the internal waves during this cast is presented in Figure 18, constructed from the differential vertical movements of the LHPR. In this scheme the maximum vertical velocities (around 0.4 m/s as stated above) are associated with transitions between trough and crest and coincide with the sectors boundaries defined in Figure 17. The distribution expected was a symmetrical accumulation of zooplankton biomass above the troughs and below the crests of the internal waves. Since the convergence-divergence processes were actually taking place during the sampling, the accumulation recorded was higher towards the end of the cast (darker accumulation areas on the right-hand side of Figure 17).

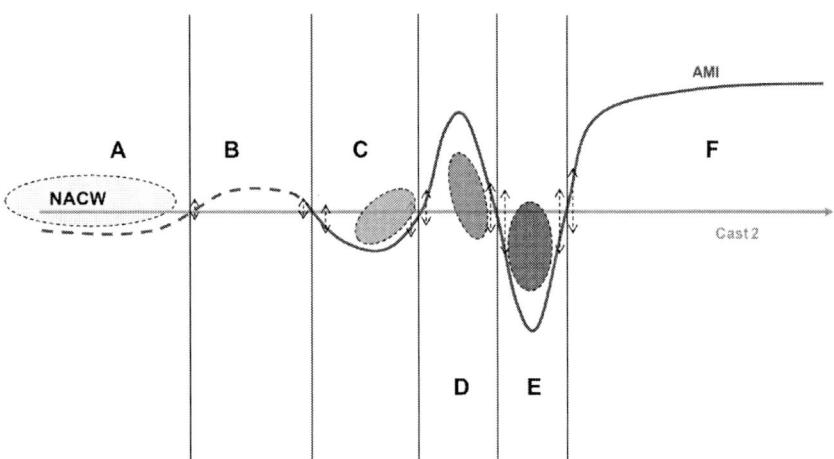

Figure 18. Schematic representation of internal wave's develop during cast 2 inferred from the vertical movement of the LHPR. Small arrows represent zones of stronger vertical movement which usually happens between crest and trough and coincides with sector limits (from figure 17). Shaded areas are preferential accumulation zones for plankton biomass.

Sector C of this final cast coincides with sector C of the previous one (see Figure 16) and is located directly over the sill (Figure 13) or slightly westward. If the divergent processes do occur, as proposed above, it is precisely in this area where they will be most likely (see Macias et al, 2007; 2008). As commented above, during the final cast a high chlorophyll concentration (Figure 17) was found in this sector, while during the previous cast this concentration was extremely low (Figure 15). The fact than only a few hours passed between one cast and the next (insufficient time for *in situ* growth), and the presence of macroalgae debris (gen. *Plocamium*) in the samples of this sector in the last cast, are two factors suggesting that the hypothesis of lateral advection of chlorophyll-rich coastal areas is very likely. This is also supported by the intermediate values of salinity and temperature recorded in this sector C (Figure 17), which match well with the presence of partially-mixed waters from the coastal and shallower sectors (Navarro et al, 2006, Macías et al, 2007). The steep horizontal gradients of temperature and salinity between sectors C and D impede the propagation of this chlorophyll patch eastward. The effect is to pack individuals together and concentrate the phytoplankton in a reduced area, making the chlorophyll concentration in the other sectors virtually zero (Figure 17).

Within this last cast sharp differences in zooplankton biomass distributions are found between sectors (Figure 16, lower panel). Maximum concentrations appear situated over the wave trough (see Figure 18) and below the wave crest (sectors C, D and E; Figure 16, lower panel) as predicted by the model of Lernnet-Cody and Franks (1999). However, not only biomass distribution is affected by the internal waves but also taxonomic composition (Munk et al, 2003). In this study, bigger (and likely better) swimming organisms appear in the samples from sectors D and E, induced by the presence of the arrested internal waves in this region, while ostracods with lower swimming velocities decrease in abundance from sectors A to E, but increase again in sector F where the effects of internal wave generation are no longer felt.

The reported data present evidence of intense coastal-channel interactions in the Camarinal Sill region, and help explaining how the combined presence of such processes with the internal waves created over the sill could create the spatio-temporal pattern of plankton biomass and its distribution (Macías et al, 2010b). This is the first time that direct evidences (i.e. quick planktonic biomass accumulation) of such interactions are obtained and give a strong support to the hypothesis presented along this entire chapter about the generation mechanisms of those processes as the timings and places of the aggregations match perfectly the theoretical interpretation exposed elsewhere.

It is also probable that these biological patterns could also be part of the explanation of the special food web structure in the pelagic realm of the Strait (Macías et al, 2010b) which presents notable differences with respect to other nearby basins (Gulf of Cádiz and Alboran Sea), as high abundance of seabirds, high-level predators and marine mammals have being reported within the Strait itself. This high abundance has been associated with the strategic position of the Strait between two marine basins and two continents (as a natural pathway for large-scale migration). However, the characteristics of its marine environment may also partially explain those particular trophic features as the accumulation of biomass of a size suitable for consumption by higher trophic levels (such as zooplankton) is of vital importance for the dynamics and structure of the marine food web, as seabirds and top-predator fish (i.e tuna, swordfish,..) have been reported to accumulate on oceanic fronts (Russell et al, 1999) following higher zooplankton abundance in such areas. Furthermore, marine mammals are usually found in systems where physical forcing permanently reinforces primary and secondary production (Fielder, 2002). As both these processes take place in the Strait, they could partially explain the abundance of all these high-trophic-level organisms.

4.2. DIRECT EVIDENCES OF COASTAL-CHANNEL-TIDALLY-INDUCED INTERACTIONS

With the accumulation of evidences about the actual mechanisms of coastal-channel interactions and taking into account its potential effects on the dynamics of the marine pelagic ecosystem of the Strait and nearby adjacent basins (see discussion above) a specifically designed observation was carried out in September 2008 during a cruise on board BO Sarmiento de Gamboa.

This experiment consisted on a water-tracer release in a coastal situation (green cross in Figure 13 and 19) from a small boat during the shift from inflowing to outflowing current at a spring tide period. The used tracer was Rhodamine-WT which is a fluorescent material not found naturally in marine environments being, thereby, easily detected by using a conventional fluorometer (Turner TD-10) equipped with specific excitation/emission filters.

The surface Rhodamine concentration was recorded from BO Sarmiento de Gamboa while following the track presented with a blue line in Figure 13.

The ship moved along this zigzag line during 18 hours, i.e. covering an entire tidal cycle. Water was collected by a surface intake (nominal depth 5 meters) and conducted to the registration chamber of the fluorometer which recorded Rhodamine-related fluorescence at one hertz rate (i.e one record per second). Data were stored at the fluorometer-physical memory and later downloaded to a compatible PC computer.

The moment of the tracer release was selected to be before the more suitable moment for the horizontal divergences appearance to allow the initial patch to be expanded a little bit and being, thereby, suitable to be suctioned towards the central channel of the Strait. The results of the water samplings were later divided into "legs", each of this leg being a complete track of the blue line showed in Figure 13.

On the other hand, numerical simulations of the dispersion of a virtual tracer were carried out using a simplified version of the recently-developed UCA-3D numerical model which is nonlinear and weakly non-hydrostatic, has a spatial high-resolution and uses finite-difference schemes and sigma-coordinate in the vertical.

Current fields provided by the model for the same dates that the experiment was conducted were used to feed-in the advection module of the modelling tool Ichthyop (Lett et al, 2008) which simulate the horizontal advection and dispersion of an inert tracer with neutral buoyancy (i.e. assimilated to a Rhodamine-marked water mass). The release of the virtual tracer was simulated at the same position and time (relative to the tidal cycle) as the real one so the results of the simulation could be directly compared to the Rhodamine data collected from the boat.

4.2.1. Field Data and Modelling Results

As stated above, the tracer was released shortly before the tidal current reversion over the Camarinal Sill (see inlet in Figure 19a). Coinciding with the release, the vessel started registering the surface Rhodamine-fluorescence along the sampling path beginning by the south-west corner (red arrow in Figure 19a). As can be observed, at the south of the sampling area no Rhodamine was detected at this stage as the recorded concentration were very close to the analytical determination threshold (situated at 0.1 ppb). While the vessel navigated on-shore, Rhodamine concentration increases progressively particularly at the western side of the medium line and in the closest region to the release area.

Simulations of the advection-diffusion model for the inert tracer shows a quite good agreement with the observations. The initial patch is moving westward when the surface current reinforces its outflowing characteristics (from 1 hour after the release). At "Punta Paloma" cape (see Figure 19) the patch is stretched in a north-south direction coinciding with the curvature of the bottom-ridge which constitutes the Camarinal Sill. This region of southward transport matches exactly with the area of higher Rhodamine concentration found in the vessel-based sample.

The second sampling leg started when the currents over the Sill changed again to east flowing (i.e. after the HW time) and begins in the north-east tip of the sampling path (see Figure 19b). Higher Rhodamine concentrations were found in the eastern side of the sampled area, corresponding to a quick advection of the marked water masses with the increasing inflowing flux. At the same time quite high concentrations were detected in the middle of the channel of the Strait (at about 35.9°N) confirming the suction of coastal waters during this tidal cycle.

The advection pattern predicted by the numerical model also agrees well with the reported data as the virtual markers were quickly moving towards the east with the increased inflowing current although the path followed by the virtual particles seems to be closer to the shore that the Rhodamine data. This second sampling leg finished almost 12 hours after the Rhodamine release so nearly one tidal cycle was already sampled so far.

Immediately after the end of this sampling, the third leg started (Figure 19c), which only lasted three hours. It was conducted from the south (i.e. from the Camarinal Sill area) towards the shore and in this case only significant Rhodamine fluorescence was found at the easternmost side of the inner line, almost in front of the town of Tarifa. The model also predicts accumulation of the virtual tracer in this same region though slightly displaced towards the east.

The final sampling leg was done from inshore and lasted also three hours (Figure 19d). In this case two separate patches of high Rhodamine concentration were detected in the inner and middle lines which could indicate that some of the water which was moving eastward during the previous leg was able to reverse its movement and come back forced by the new change on the tidal current direction over the Camarinal Sill (see inlet in Figure 19d). This effect is not well reproduced by the model which predicts that the virtual tracer keep travelling eastward during this period.

This lack of agreement in the last sampling leg could be attributable to the slightly quicker eastward advection of the marked water masses predicted by the model. As already pointed out, during the third leg almost all the virtual particles were found to the east of Tarifa point while some of the Rhodamine still remains westwards. It is already known that surface-tidally-induced reversion only happens westwards of Tarifa (see chapter 3 of the book) so this could explain why the model does not reproduce the reversion of the marked water masses observed in the field data.

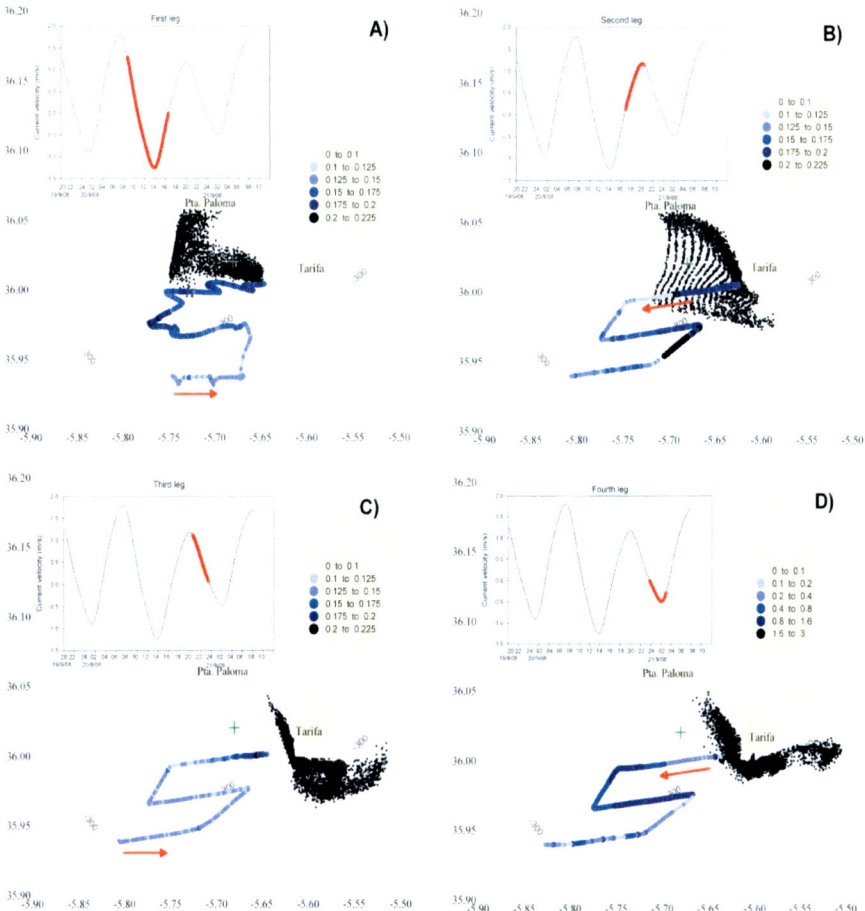

Figure 19. Rhodamine concentration in the surface waters during the different sampling legs (blue dots). Predicted dispersion of the virtual tracer by the hydrological model (black dots). Red arrow marks the travel direction of the vessel when following the sampling path showed in figure 13. Inlets show the tidal current during each of the sampling legs.

However, and in spite of this slight incongruence, both model and field data clearly show that coastal waters are effectively transported towards the main channel of the Strait during the east flowing phase of the tidal cycle. Both dataset also points to the region south of "Punta Paloma" cape as the most likely to the offshore suction of the water masses.

This same region has been identified as very relevant for these processes by Vázquez et al (2009) in a very recent study combining Advanced Synthetic Aperture Radar (ASAR Image Mode) images and surface color imagery (MODIS and MERIS sensor). These authors use the remote information from the different sensors to study the generation (and propagation) of several internal waves episodes over the Camarinal Sill and associated 2D surface structures and relate them to the tidal conditions within the Strait. These are the first direct evidences of such interactions which confirm the proposed hypothesis and mechanisms already exposed above

As a repetitive pattern in all the internal waves generation processes studied, a tongue of high chlorophyll waters appears extending southward in the region of the "Punta Paloma" cape, thereby, confirming the findings of the tracer-release experiment and the model simulations showed here.

Chapter 5

CONCLUSION

The main conclusion which could be outdrawn of this entire chapter is that tidal dynamics control and shapes as the main agent the general hydrodynamics of the Strait of Gibraltar. In this sense it is necessary to acknowledge that other factors such as winds (e.g. Macías et al., 2008b) or atmospheric pressure distributions (e.g. Vázquez et al, 2008) could also play a role in controlling the hydrography of the Strait and nearby areas but at a much lower level that the tide does.

As usually happens in the marine environment, a strong coupling between physical forcing and biogeochemical patterns is found in the Strait with the main biological processes being shaped and controlled by hydrological factors. The interaction mechanisms found in the area include:

i. A strong interfacial mixing (between Mediterranean and Atlantic waters) take place along the main channel of the Strait giving rise to the development of a intermittent upwelling with a tidal-related periodicity. This process injects nutrients in the upper layer of the incoming Atlantic Jet and not only alters the biogeochemical budget between both basins but also should have consequences on the dynamics of the pelagic ecosystem of the nearby Alboran Sea.

ii. The strong advection associated to the baroclinic and barotropic tidally-related along-strait currents prevents the effect of this upper-layer fertilization to take place within the Strait itself but rather makes it more suitable to be used (the injected nutrients) in the pelagic ecosystem of the eastern Alboran Sea.

iii. Complex water-masses circulation pattern through the Strait are induced by the tidal dynamics and associated current fields. This creates a discontinuous presence of different biomass accumulations with specific characteristics and particular origins.
iv. There are coastal-channel pulsating interactions driven by the horizontal surface divergence induced by the tidal-currents reversion. This makes coastal material to be injected within the main along-strait flux into open regions of the Alboran Sea.

REFERENCES

Alonso del Rosario, J.J., Bruno Mejías, M., Vázquez-Escobar, A. (2003). The influence of tidal hydrodinamic conditions on the generation of lee waves at the main sill of the Strait of Gibraltar. *Deep-Sea Research* I, 50, 1005-1021.

Armi, L., Farmer, D. (1988). The flow of Mediterranean Water through the Strait of Gibraltar. *Prog. Oceanog.,* 21, 41-82.

Arnone, R.A., Wiesenburg, D.A., Saunders, K.D. (1990). The Origin and Characteristics of the Algerian Current. *Journal of Geophysical Research,* 95, 1587-1598.

Basheck, B., Send, U., García Lafuente, J., Candela, J. (2001). Transport estimates in the Strait of Gibraltar with a tidal inverse model. *Journal of Geophysical Research,* 106, 31033-31044.

Béthoux, J.P., Copin-Montégut, G. (1986). Biological fixation of atmospheric nitrogen in the Mediterranean Sea. *Limnol. Oceanogr.,* 31(6): 1353-1358.

Béthoux, J.P., Morin, P., Chaumery, C., Connan, O., Gentili, B., Ruiz-Pino, D. (1998). Nutrients in the Mediterranean Sea, mass balance and statistical analysis of concentrations with respect to environmental change. *Marine Chemistry,* 63, 155-169.

Bone, Q. (2005). Gelatinous animals and physiology. *Journal of Marine Biology Association of U.K.,* 85, 641-653.

Boyce, F.M. (1975). Internal Waves in the Strait of Gibraltar. *Deep-Sea Research,* 22, 597-610.

Boyra, G., Irigoien, X. and Arregi, I. (2005). Plankton visual analyzer. *Globec International Newsletter,* April.

Brandt, P., Alpers, W., Backhaus, J.O. (1996). Study of the generation and propagation of internal waves in the Strait of Gibraltar using a numerical model and synthetic aperture radar images of the European ERS 1 satellite. *Journal of Geophysical Research,* 101, 14237-14252.

Bray, N.A., Ochoa, J., Kinder, T.H. (1995). The role of interface in exchange through the Strait of Gibraltar. *Journal of Geophysical Research,* 100, 10755-10776.

Briscoe, M.G. (1984). Tides, solitons and nutrients. *Nature,* 312, 15-17.

Bruno, M., Alonso, J.J., Cózar, A., Vidal, J., Ruiz-Cañavate A., Echevarría, F., Ruiz, J. (2002). The boiling-water phenomena at Camarinal Sill, the strait of Gibraltar. *Deep-Sea Research* II, 49, 4097-4113.

Bryden, H.L., Kinder, T.H. (1988). Gibraltar experiment: a plan for dynamic and kinematic investigation of strait mixing, exchange and turbulence. *Oceanologica Acta,* N_SP.

Bryden, H.L., Kinder, T.H. (1991). Steady two-layer exchange through the Strait of Gibraltar. *Deep-Sea Research,* 38, 445-463.

Bryden, H.L., Candela, J., Kinder, T.H. (1994). Exchange through the Strait of Gibraltar. *Progress in Oceanography,* 33, 201-248.

Candela, J., Winant, C., Bryden, H. (1989). Meteorologically Forced Subinertial Flows Through the Strait of Gibraltar. *Journal of Geophysical Research,* 94, 12667-12679.

Candela, J. (1990). The barotrophic tide in the Strait of Gibraltar. In: K.A. Publisher (Editor), *The Physical Oceanography of Sea Straits,* pp. 457-475.

Candela, J., Winant, C., Ruiz, A. (1990). Tides in the Strait of Gibraltar. *Journal of Geophyisical Research- Oceans,* 95, 7313-7335.

Candela, J. (1991). The Strait of Gibraltar and its role in the dynamics of the Mediterran Sea. *Dynamics of Atmospheres and Oceans,* 15, 267-299.

Castro, M.J., García-Rodriguez, J.A., González-Vida, J.M., Macías,J., Parés, C., Vázquez-Cendón, M.E. (2004). Numerical simulation of two-layer shallow water flows through channels with irregular geometry. *Journal of Computational Physics,* 195, 202-235.

Chisholm, S.W. (1992). Phytoplankton size. In: P.G. Falkowski, Woodhead, A.D. (Editor), *Primary productivity and Biogeochemical cycles in the Sea.* Plennun Press, New York, pp. 213-237.

Corzo, A., Morillo, J.A., Rodríguez, S. (2000). Production of transparent exopolymer particles (TEP) in cultures of Chaetoceros calcitrans under nitrogen limitation. *Aquatic Microbial Ecology,* 23, 63-72.

Crepon, M. (1965). Influence de la pression atmospherique sur le niveau moyen de la Mediterranee Occidentale et sur le flux a travers le detroit de Gibraltar. *Cahiers Oceanographiques,* 1, 15-32.

Dafner, E.V., Boscolo, R., Bryden, H.L. (2003). The N:Si:P molar ratio in the Strait of Gibraltar. *Geophysical Research Letters*, 30, 13.1-13.4.

Echevarria, F., García-Lafuente, J.,Bruno, M.,Gorsky, G, Goutx, M.,González, N.,García, C.M.,Gómez, F.,Vargas, J.M.,Picheral, M.,Striby, L.,Varela, M.,Alonso, J.J.,Reul, A., Cózar, A.,Prieto, L.,Sarhan, T.,Plaza, F.,Jiménez-Gómez, F.(2002). Physical-biological coupling in the Strait of Gibraltar. *Deep-Sea Research II,* 49, 4115-4130.

Echevarria, F., Zabala, L., Corzo, A., Navarro, G., Prieto, L., Macías, D., (2009). Spatial distribution of autotrophic picoplankton in relation to physical forcings: the Gulf of Cádiz, Strait of Gibraltar and Alborán Sea case study. *Journal of Plankton Research,* 33, 1339-1351.

Elbaz-Poulichet, F., Morley, N.H., Beckers, J.M., Nomerange, P. (2001). Metal fluxes through the Strait of Gibraltar: the influence of the Tinto and Odiel rivers (SW Spain). *Marine Chemistry,* 73, 193-213

Eppley, R.W., Rogers, J.N., McCarthy, J.J. (1969). Half saturation constrats for uptake of nitrate and ammonium by marine phytoplankton. *Limnology and Oceanography,* 14, 912-920.

Estrada, M., Marrase, C., Latasa, M., Berdalet, E., Delgado, M., Riera, T. (1993). Variability of the Deep Chlorophyll Maximum characteristics in the northwestern Mediterranean. *Marine Ecology Progress Series*, 92, 289-300.

Farmer, D., Armi, L. (1986). Maximal two-layer exchange over a sill and through the combination of a sill and contraction with barotropic flow. *Journal of Fluid Mechanic*, 164, 53-76.

Fasham, M.J.R., Ducklow, H.W., McKelvie, S.M. (1990). A nitrogen-based model of plankton dynamics in the oceanic mixed layer. *Journal of Marine Research,* 48, 591-639.

Fielder, P.C. (2002). The annual cycle and biological effects of the Costa Rica Dome. *Deep-Sea Research I,* 49, 321-338.

Franks, P. (1992). Sink or swim: accumulation of biomass at fronts. *Marine Ecology Progress Series,* 82, 1-12.

García-Lafuente, J., Vargas, J.M., Plaza, F., Sarham, T., Candela, J., Basheck, B. (2000). Tide at the eastern section of the Strait of Gibraltar. *Journal of Geophysical Research,* 105, 14197-14213.

García-Lafuente, J., Criado, F. (2001). La climatología y la topografía del Estrecho de Gibraltar determinantes de las propiedades termohalinas del agua del Mar Mediterráneo., *Física de la Tierra,*13, pp. 43-54 (in Spanish).

García-Lafuente, J., Delgado, J., Vargas, J.M., Vargas, M., Plaza, F., Sarhan, T. (2002). Low frequency variability of the exchanged flows through the Strait of Gibraltar during CANIGO. *Deep-Sea Research II*, 49, 4051-4067.

García-Lafuente, J., Vargas Domínguez, J.M. (2003). Recent observations of the exchanged flows through the Strait of Gibraltar and their fluctuations at different time scales. *Recent Research Development in Geophysics*, 5, 73-84.

García-Lafuente, J., Ruiz, J. (2007). The Gulf of Cádiz pelagic ecosystem. *Progress in Oceanography*, 74, 228-251.

García-Lafuente, J., Sánchez Román, A., Díaz del Río, G., Sannino, G., Sánchez Garrido, J.C. (2007). Recent observations of seasonal variability of the Mediterranean outflow in the Strait of Gibraltar. *Journal of Geophysical Research*, 112, C10005.

García-Lafuente, J. (2008). Gulf of Cádiz and Strait of Gibraltar in The Seas of Spain. Ministerio de Medio Ambiente y Medio Rural y Marino, 508 pp.

Gascard, J.C., Richez, C. (1985). Water Masses and Circulation in the Western Alboran Sea and in the Straits of Gibraltar. *Progress in Oceanography*, 15, 157-216.

Goericke, R., Olson, R.J., Shalapyonok, A. (2000). A novel niche for Proclorococcus sp. in low ligth suboxic environments in the Arabian Sea and the Eastern Tropical North Atlantic. *Deep-Sea Research* I, 47, 1183-1205.

Gómez, F., Echevarría, F., García, C.M., Prieto, L., Ruíz, J., Reul, A., Jiménez-Gómez, F., Varela, M. (2000a). Microplankton distribution in the Strait of Gibraltar: coupling between organism and hydrodinamics structures. *Journal of Plankton Research*, 22, 603-617.

Gómez, F., González, N., Echevarría, F., García, C.M. (2000b). Distribution and fluxes of dissolved nutrients in the Strait of Gibraltar and its relationships to microphytoplankton biomass. *Estuarine, Coastal and Shelf Science*, 51, 439-449.

Gómez, F., Gorsky, G.,Striby, L.,Vargas, J.M.,Gónzalez, N.,Picheral, M.,García-Lafuente, J.,Varela, M.,Goutx, M. (2001). Small-scale temporal variations in biogeochemical features in the Strait of Gibraltar, Mediterranean side- the role of NACW and the interface oscillation. *Journal of Marine Systems*, 30, 207-220.

Gómez, F., Gorsky, G.,García-Górriz, E., Picheral, M. (2004). Control of the phytoplankton distribution in the Strait of Gibraltar by wind and fortnightly tides. *Estuarine, Coastal and Shelf Science*, 59: 485-497.

Gran, H.H., 1931. On the conditions for the production of plankton in the sea. *Raap. P.-v Réun. Cons. int. Explor. Mer.,* 75, 37-46.

Herbland, A., Voituriez,B. (1979). Hydrological structure analysis for estimating the primary production in the tropical Atlantic Ocean. *Journal of Marine Research,* 37: 87-101.

Hopkins, T.S. (1999). The thermohaline forcing of the Gibraltar exchange. *Journal of Marine Systems,* 20, 1-31.

Huertas, E., Navarro, G., Rodríguez-Gálvez, S., Prieto, L. (2005). The influence of phytoplankton biomass on the spatial distribution of carbon dioxide in surface sea water of a coastal area of the Gulf of Cádiz (southwestern Spain). *Canadian Journal of Botany,* 83, 929-940.

Izquierdo, A., Tejedor, L., Sein, D.V., Backhaus, J.O., Brandt, P., Rubino, A., Kagan, B.A. (2001). Control Variability and Internal Bore Evolution in the Strait of Gibraltar: A 2-D Two-Layer Model Study. *Estuarine, Coastal and Shelf Science,* 53, 637-651.

Krzanowski, W.J. (1998). Principles of Multivariate Análisis. *A user's Perspective.* Oxford Press.

La Violette, P.E., Arnone, R.A. (1988). A tide-generated internal waveform in the western approaches to the Strait of Gibraltar. *Journal of Geophysical Research,* 93, 15653-15667.

Lacombe, H., Richez, C. (1982). The regime of the Strait of Gibraltar. In: J.C.J. Nilhoul (Editor), Elsevier Oceanography Series. 34.

Le Provost, C., Lyard, F., Molines, J. M., Genko, M. L. & Rabillloud, F. (1998). A hydrodynamic ocean tide model improved by assimilating a satellite altimeter-derived data set. *Journal of Geophyisical Research,* 103, 5513-5529.

Leet, C., Verley, P. , Mullon, C., Parada, C., Brochier, T., Penven, P., Blanke, B. (2008). A lagrangian tool for modelling ichtyoplankton dynamics. *Environmental Modelling & Software,* 23, 1210-1214.

Lennert-Cody, C.E., Franks, P.J.S. (1999). Plankton patchiness in high-frequency internal waves. *Marine Ecology Progress Series,* 186, 59-66.

Li, W.K.W. (2002). Macroecological patterns of phytoplankton in the northwestern North Atlantic Ocean. *Nature,* 419, 154-157.

Longhurst, A., Harrison, A. (1989). The biological pump: profiles of plankton production and consumption in the upper ocean. *Progress in Oceanography,* 22, 47-123.

Macías, D., García, C.M., Echevarría, F., Vázquez-Escobar, A., Bruno, M. (2006). Tidal induced variability of mixing processes on Camarinal Sill

(Strait of Gibraltar).A pulsating event. *Journal of Marine Systems,* 60, 177-192.

Macías, D., Martin, A.P., García Lafuente, J., García, C.M.,Yool, A., Bruno, M.,Vázquez, A., Izquierdo, A., Sein, D., Echevarría, F. (2007). Mixing and biogeochemical effects induced by tides on the Atlantic-Mediterranean flow in the Strait of Gibraltar. An analysis through a physical-biological coupled model. *Progress in Oceanography,* 74, 252-272.

Macías, D., Martin, A.P., García Lafuente, J., García, C.M.,Yool, A. Bruno, M.,Vázquez, A., Izquierdo, A., Sein, D., Echevarría, F. (2007[a]). Mixing and biogeochemical effects induced by tides at the Strait of Gibraltar: a modelling study. *Globec International Newsletter,* 13(2), 25-27.

Macías, D., Lubian, L.M., Echevarría, F., Huertas, E., García, C.M. (2008). Chlorophyll maxima and water mass interfaces: tidally induced dynamics in the Strait of Gibraltar. *Deep-Sea Research I,* 55, 832-846.

Macías, D., NAvarro, G., Bartual, A., Echevarría, F., Huertas, I.E. (2009). Primary production in the Strait of Gibraltar: carbon fixation rates in relation to hydrodynamic and phytoplankton dynamics. *Estuarine Coastal and Shelf Science,* 83, 197-210.

Macías, D., Ramírez, E., García, C.M. (2010a). Effect of nutrient input frequency on the structure and dynamics of the marine pelagic community. A modelling approach. *Journal of Marine Research.68(2)*

Macías, D., Somavilla, R., González-Gordillo, I., Echevarría, F. (2010b). Physical control on zooplankton distribution pattern at the Strait of Gibraltar during an episode of internal wave generation. *Marine Ecology Progress Series*, 408:79-95.

Malone, T.C. (1980). Algal size. In: I. Morris (Editor), The Physological Ecology of Phytoplankton. *Blackwell Scientific Publications,* Oxford, pp. 133-463.

Mann, K.H., Lazier, J.R.N. (1991). Dynamics of marine ecosystems. Biological-Physical Interactions in the Oceans. *Blackwell Scientific Publications.*

Margalef, R. (1989). *Ecología. Ediciones Omega.* (in Spanish).

Mercado, J.M., Ramírez, T., Cortés, D., Sebastián, M., Vargas-Yáñez, M. (2005). Seasonal and inter-annual variability of the phytoplankton communities in an upwelling area of the Alborán Sea (SW Mediterranean Sea). *Scientia Marina,* 69, 451-465.

Minas, H.J., Coste, B., Le Corre, P., Minas, M., Raimbault, P. (1991). Biological and Geochemical Signatures Associated With the Water

Circulation Through the Strait of Gibraltar and in the Western Alboran Sea. *Journal of Geophyisical Research,* 96, 8755-8771.

Minas, H.J., Minas, M. (1993). Influence du Détroit de Gibraltar sue la biogéochime de la Méditerranée et du proceh Atlantique. *Annales de L'Institute Océanographique,* 69, 203-214.

Morán, X.A.G., Taupier-Letage, I., Vázquez-Dominguez, E., Ruiz, S., Arin, L., Raimbault, P., Estrada, M. (2001). Physical-biological coupling in Algerian Basin (SW Mediterranean): Influence of mesoscales instabilities on the biomass and production of phytoplankton and bacterioplankton. *Deep-Sea Research I,* 48, 405-437.

Munk, P., Hansen, B.W., Nielsen, T.G., Thomsen, H.A. (2003). Changes in plankton and fish larvae communities across hydrographic fronts off West Greenland. *Journal of Plankton Research,* 25, 815-830.

Mutlu, E. (2006). Diel vertical migration of Sagitta setosa as inferred acoustically in the Black Sea. *Marine Biology,* 149, 573-584.

Navarro, G. (2004). Escalas de variación espacio-temporal de procesos pelágicos en el Golfo de Cádiz, PhD Thesis. Universidad de Cádiz, 207 pp (in Spanish).

Navarro, G., Ruiz, J., Huertas, I.E., García, C.M., Criado-Aldeanueva, F., Echevarría, F. (2006). Basin scale structures governing the position of the subsurface chlorophyll maximum in the gulf of Cádiz. *Deep-Sea Research II,* 53, 1261-1281.

Ochoa, J., Bray, N.A. (1991). Water mass exchange in the Gulf of Cadiz. *Deep Sea Research,* 38, S465-S503.

Packard, T.T., Minas, H.J., Coste, B., Martinez, R., Bonin, M.C., Gostan, J., Garfield, P., Christensen, J., Dortch, Q., Minas, M., Copin-Montegut, G., Copin-Montegut, C. (1988). Formation of the Alboran oxygen minimun zone. *Deep-Sea Research,* 35, 1111-1118.

Pan, L.A., Zhang, L.H., Zhang, J., Gasol, J.M., Chao, M. (2004). On board flow cytometry observations of picoplankton comunity structure in the East China Sea during the fall of different years. *FEMS Microbiology Ecology,* 52, 243-253.

Pettigrew, N.R. (1989). Direct measurements of the Flow of Western Mediterranean Deep Water Over the Gibraltar Sill. *Journal of Geophysical Research,* 94, 18089-18093.

Prieto, L., G. Navarro, A. Cózar, F. Echevarría, C.M. García (2006). Distribution of TEP in the euphotic and upper mesopelagic zones of the southern Iberian coasts. *Deep-Sea Research II,* 53, 1314-1328.

Reul, A., Vargas, J.M., Jiménez-Gómez, F., Echevarría, F., García-Lafuente, J., Rodriguez,J. (2002). Exchange of planktonic biomass through the Strait of Gibraltar in late summer conditions. *Deep-Sea Research* II, 49, 4131-4144.

Reul, A., Muñoz, M., Criado-Aldeanueva, F., Rodriguez, V. (2006). Spatial distribution of phytoplankton<13 μm in the Gulf of Cadiz in relation to water masses and circulation pattern under westerlies and easterlies wind regimes. *Deep-Sea Research* II, 53, 1294-1313.

Richez, C. (1994). Airbone synthetic-aperture radar tracking of internal waves in the Strait of Gibraltar. *Progress in Oceanography*, 33, 93-101.

Rodríguez, J., Blanco, J.M. , Jiménez-Gómez, F., Echevarría, F., Gil, J., Rodríguez, V., Ruiz, J., Bautista, B., Guerrero, F. (1998). Patterns in the size structure of the phytoplankton community in the deep fluorescence maximum of the Alboran Sea (southwestern Mediterranean). *Deep-Sea Research* I, 45, 1577-1593.

Ruiz, J., Echevarría, F., Font, J., Ruiz, S., García, E., Blanco, J.M., Jiménez-Gómez, F., Prieto, L., González-Alaminos, A., García, C.M., Cipollini, P., Snaith, H., Bartual, A., Reul, A., Rodríguez, V. (2001). Surface distribution of chlorophyll, particles and gelbstoff in the Atlantic jet of the Alborán Sea: from submesoscale to subinertial scales of variability. *Journal of Marine Systems*, 29, 277-292.

Ruiz, J., Macías, D., Peters, F. (2004). Turbulence increases the average settling velocity of phytoplankton cells. *PNAS*, 101, 17720-17724.

Russell, R.W., Harrison,N. M. and Hunt Jr.,G. L. (1999). Foraging at a front: Hydrography, zooplankton, and avian planktivory in the northern Bering Sea. *Marine Ecology Progress Series*, 182, 77-93.

Sannino, G., Bargagli, A., Artale, V. (2002). Numerical model of the mean exchange through the Strait of Gibraltar. *Journal of Geophysical Research*, 107, 3044.

Sannino, G., A. Bargagli, Artale, V. (2004). Numerical modeling of the semidiurnal tidal exchange through the Strait of Gibraltar. *Journal of Geophyisical Research*, 109, C05011.

Sánchez-Garrido, J.C., García-Lafuente, J., Criado Aldeanueva, F., Baquerizo, A., Sannino, G.(2008). Time-spatial variability observed in velocity of propagation of the internal bore in the Strait of Gibraltar. *Journal of Geophyisical Research*, 113, C07034.

Sein, D.V., Backhaus, J.O., Brandt, P., Izquierdo, A., Kagan, B.A., Rubino, A., Tejedor, L. (1998). Flow exchange and tidally induced dynamics in the Strait of Gibraltar as derived from a two-layer, boundary-fitted

coordinated model. Unesco Workshop on Ocenic Fronts and Related Phenomena.
Smetacek, V., Passow, U. (1990). Spring bloom initiation and Sverdrup's critical-depth model. *Limnology and Oceanography,* 35, 228-234.
Sverdrup, H.U. (1953). On conditions for the vernal blooming of phytoplankton. *J. Cons. Perm. Int. Exp. Mer.,* 18, 287-295.
Tsimplis, M.N., Bryden, H.L. (2000). Estimation of the transports through the Strait of Gibraltar. *Deep-Sea Research.* I, 47, 2219-2242.
van Genn, A., Rosener, P., Boyle, E. (1988). Entrainment of trace-metal-enriched-Atlantic-shelf water in the inflow to the Mediterranean Sea. *Nature,* 331, 423-426.
Vargas, J.M., García-Lafuente,J., Candela, J., Sánchez, A.J. (2006). Fortnightly and monthly variability of the exchange through the Strait of Gibraltar. *Prog. Oceanog.,* 70, 466-485.
Vázquez, A., Stashchuk, N., Vlasenko, V., Bruno, M., Izquierdo, A., Gallacher, P.C. (2006). Evidence of multimodal structure of the baroclinic tide in the Strait of Gibraltar. *Geophysical Research Letters,* 33, L17605.
Vázquez, A., Bruno, M., Izquierdo, Macías, D. (2008). The effect of meteorologically forced subinertial flows on internal waves generation at the main sill of the Strait of Gibraltar. *Deep-Sea Research I,* 57, 1277-1283.
Vázquez, A., Flecha, S., Bruno, M., Macías, D., Navarro, G. (2009). Internal waves and short-scale distribution patterns of chlorophyll in the Strait of Gibraltar and Alborán Sea. *Geophysical Research Letters,* 36, L23601.
Wang, D.P. (1989). Model of mean and tidal flows in the Strait of Gibraltar. *Deep-Sea Research,* 36, 1535-1548.
Wang, D.P. (1993). The Strait of Gibraltar Model: Internal tide, diurnal inequality and fortnightly modulations. *Deep-Sea Research I,* 40A, 1187-1203.
Wash, J.J. (1988). On the nature of continental shelves. Academic Press. New York, 126.
Wesson, J.C., Gregg, M.C. (1994). Mixing at Camarinal Sill in the Strait of Gibraltar. *Journal of Geophysical Research,* 99, 9847-9878.
Wiebe, P.H., Benfield, M.C. (2003). From the Hensen net toward a four-dimensional biological oceanography. *Progress in Oceanography,* 56, 7-136.
Wu, P., Haines, K. (1996). Modelling the dispersal of Levantine Intermediate Water and its role in Mediterranean deep water formation. *Journal of Geophysical Research,* 101, 6591-6607.

Yamasaki, H., Squires, K.D. (1996). Comparison of oceanic turbulence and copepod swimming. *Marine Ecology Progress Series,* 144, 299-301.

Zar, J.H. (1974). Biostatistical analysis. Prentice Hall International Editions.

INDEX

A

accessibility, 1
accounting, 44
Africa, 1
aggregates, 32
aggregation, 38
alternative hypothesis, 27
alters, 53
ammonium, 57
amplitude, vii, 4, 5, 7, 8, 10, 11, 12, 17, 23, 24, 32, 35, 45
atmospheric pressure, 5, 7, 10, 53
authors, 10, 13, 35, 52

B

bacteria, 17
behavior, 13
biological processes, 53
biomass, 13, 14, 29, 35, 37, 40, 42, 44, 45, 46, 47, 48, 54, 57, 58, 59, 61, 62
birds, 1

C

carbon, 23, 29, 59, 60
case study, 57
cast, 18, 35, 38, 40, 41, 42, 43, 44, 45, 46, 47
cell, 25
channels, 56
character, 24
China, 61
chlorophyll, 4, 18, 19, 20, 25, 26, 28, 29, 30, 31, 32, 33, 36, 40, 41, 42, 43, 44, 45, 47, 52, 61, 62, 63
circulation, vii, 1, 2, 3, 11, 12, 13, 14, 15, 24, 32, 33, 35, 54, 62
color, 52
communication, 1
community, 4, 13, 35, 37, 38, 40, 44, 60, 62
compensation, 28
complexity, 13, 26
components, 5, 7
composition, 4, 31, 32, 35, 37, 38, 40, 44, 47
computing, 24, 38
concentration, 4, 12, 16, 20, 23, 24, 25, 26, 28, 29, 30, 31, 37, 40, 42, 44, 45, 47, 48, 49, 50, 51
conceptual model, 13, 14, 18
concordance, 27
confidence, 29, 30
confidence interval, 29, 30
conservation, 21
consumption, 48, 59
contradiction, 28
control, vii, 53, 60

convergence, 28, 46
correlation, 12, 24, 25, 29
correlation coefficient, 25, 29
Costa Rica, 57
coupling, 19, 29, 53, 57, 58, 61
cycles, 7, 12, 14, 16, 17, 28, 56
cytometry, 61

D

data set, 59
density, 5, 6, 7, 9, 15, 16, 32, 44
diffusion, 50
direct measure, 3
discharges, 2, 28
discriminant analysis, 40
dispersion, 8, 25, 49, 51
displacement, 7, 19, 44
distribution, 4, 13, 14, 15, 17, 18, 21, 25, 29, 31, 35, 36, 37, 38, 40, 45, 46, 47, 57, 58, 59, 60, 62, 63
divergence, 27, 28, 32, 38, 45, 46, 54
diversity, 4
drainage, 1
dynamics, 2, 3, 4, 11, 13, 15, 18, 20, 25, 26, 27, 29, 35, 37, 45, 48, 53, 54, 56, 57, 59, 60, 62

E

ecosystem, 4, 13, 17, 18, 35, 37, 48, 53, 58
emission, 48
energy, 35
environment, 26, 45
environmental change, 55
estimating, 59
evaporation, 6
evolution, 14, 18, 19, 20, 26
excitation, 48
exercise, 21

F

fertilization, 53

filters, 48
fish, 48, 61
fixation, 55, 60
flank, 9
flood, 10, 21
flooding, 8
fluctuations, 5, 7, 10, 11, 58
fluorescence, 4, 17, 18, 38, 49, 50, 62
focusing, 11
food, 48
freshwater, 2

G

generation, 9, 10, 11, 31, 35, 37, 38, 39, 47, 48, 52, 55, 60, 63
groups, 35, 44
growth, 8, 25, 26, 27, 31, 47

H

height, 3, 23, 39
hypothesis, 24, 30, 31, 45, 47, 52

I

imagery, 52
images, 52, 56
in vivo, 4
income, 2
inequality, 10, 63
inhibition, 10
initiation, 63
interaction, 3, 5, 8, 13, 28, 37, 46, 53
interactions, 4, 24, 37, 38, 45, 47, 48, 52, 54
interface, 6, 7, 8, 12, 16, 21, 28, 29, 30, 31, 32, 33, 34, 35, 56, 58
interval, 17
isotherms, 19

L

liberation, 39

limitation, 56
line, 1, 2, 13, 14, 16, 20, 21, 23, 26, 33, 39, 48, 49, 50
links, 7
Lion, 6
lying, 34

M

macroalgae, 47
marine environment, 48, 53
measurement, 40
measures, 38
Mediterranean, vii, 1, 2, 3, 4, 5, 6, 7, 8, 10, 11, 12, 15, 17, 18, 19, 22, 23, 24, 28, 34, 53, 55, 57, 58, 60, 61, 62, 63
memory, 49
metals, 28
migration, 48, 61
mixing, 3, 4, 6, 7, 10, 11, 12, 13, 16, 18, 19, 20, 21, 22, 24, 25, 26, 27, 28, 35, 37, 53, 56, 59
model, 9, 10, 13, 14, 15, 16, 17, 20, 21, 22, 24, 25, 26, 27, 28, 45, 47, 49, 50, 51, 52, 55, 56, 57, 59, 60, 62, 63
modeling, 62
modelling, 13, 26, 37, 49, 59, 60
models, 3, 12, 13, 35, 38
modulations, 63
movement, 46, 50
multiple factors, 3

N

Nile, 11
nitrogen, 16, 22, 23, 55, 56, 57
nutrients, 3, 16, 23, 24, 25, 26, 31, 35, 53, 56, 58

O

observations, 4, 14, 15, 17, 18, 25, 26, 28, 29, 30, 38, 50, 58, 61
order, 11, 12, 13, 14, 16, 20, 22, 25

organism, 58
oscillation, 7, 44, 46, 58
oxygen, 61

P

packaging, 28
parameter, 16
particles, 32, 50, 51, 56, 62
performance, 17, 37
periodicity, 5, 21, 25, 53
physical environment, 21
physiology, 55
phytoplankton, 13, 15, 16, 24, 25, 26, 27, 28, 29, 30, 31, 35, 45, 47, 57, 58, 59, 60, 61, 62, 63
poor, 2
population, 25, 31, 45
positive correlation, 4, 12
positive relation, 24
prediction, 18, 19
pressure, 5, 7, 38
probe, 17
production, 3, 15, 31, 35, 48, 59, 60, 61
productivity, 56
proliferation, 30
propagation, 32, 47, 52, 55, 62
properties, vii, 16, 18, 21, 29
PVA, 38, 40

R

radar, 56, 62
radiation, 16
rainfall, 2
range, vii, 7, 10, 20, 24
reason, 6, 25
recommendations, iv
rectification, 12
reflection, 12
region, 3, 4, 13, 16, 17, 21, 27, 31, 32, 34, 47, 49, 50, 52
relationship, 15, 29
reliability, 16

resolution, 13, 14, 49
respect, 7, 44, 48, 55
returns, 6

S

salinity, 6, 17, 18, 19, 20, 21, 22, 25, 26, 28, 38, 40, 41, 42, 43, 44, 47
sampling, 12, 14, 18, 19, 20, 24, 26, 31, 38, 39, 40, 41, 42, 43, 44, 45, 46, 49, 50, 51
satellite, 56, 59
saturation, 57
sea level, 6, 7
sedimentation, 32
segregation, 4
sensors, 52
shear, 7, 16, 21
simulation, 13, 22, 23, 38, 49, 56
software, 38, 40
solitons, 56
Spain, 57, 58, 59
species, 1
speed, 18, 45
Spring, 63
strategic position, 48
strategy, 12, 42, 46
stratification, 5, 8
strength, 12
summer, 6, 62
supply, 35
surface layer, vii, 27, 31, 34, 37
surface structure, 52

T

temperature, 17, 18, 19, 38, 40, 42, 44, 47
threshold, 30, 31, 49

tides, 3, 7, 8, 10, 11, 12, 17, 21, 24, 58, 60
time resolution, 17
timing, 12
tracking, 62
transition, 29, 40, 44
transitions, 46
transport, 4, 35, 50
turbulence, 3, 32, 56, 64

U

uncertainty, 21
uniform, 16, 40

V

variability, 6, 8, 11, 13, 21, 35, 38, 58, 59, 60, 62, 63
variables, 13, 15, 16, 18, 40
variations, 7, 10, 15, 16, 58
velocity, 16, 18, 19, 21, 27, 32, 34, 39, 41, 44, 46, 62
vessels, 17

W

web, 35, 48
wind, 4, 17, 58, 62
winter, 6

Z

zooplankton, 16, 38, 40, 44, 45, 46, 47, 48, 60, 62